Process Mining Techniques for Managing and Improving Healthcare Systems

This book discusses a new process mining method along with a detailed comparison between different techniques that provide a complete vision of the process of data acquisition, data analysis, and data prediction.

Process Mining Techniques for Managing and Improving Healthcare Systems offers a new framework for process learning which is probabilistic and enables the process to be learned in an accumulative manner. The steps of prediction modeling and building the required knowledge are highlighted throughout the book, along with a strong emphasis on the correlation between the healthcare domain and technology including the different aspects, such as: managing records, information, and procedures; early detection of diseases; and the improvement of accuracy in choosing the right treatment procedures.

This reference provides a wealth of knowledge for practitioners, researchers, and students at the basic and intermediary levels working within the healthcare system, computer science, electronics and communications, as well as medical providers and also hospital management entities.

Process Mining Techniques for Managing and Improving Healthcare Systems

Maha Zayoud

CRC Press
Taylor & Francis Group
Boca Raton London New York

CRC Press is an imprint of the
Taylor & Francis Group, an **informa** business

Designed cover image: Shutterstock

First edition published 2023
by CRC Press
6000 Broken Sound Parkway NW, Suite 300, Boca Raton, FL 33487-2742

and by CRC Press
4 Park Square, Milton Park, Abingdon, Oxon, OX14 4RN

CRC Press is an imprint of Taylor & Francis Group, LLC

© 2023 Maha Zayoud

ISBN: 978-1-032-43109-3 (hbk)
ISBN: 978-1-032-43275-5 (pbk)
ISBN: 978-1-003-36657-7 (ebk)

DOI: 10.1201/9781003366577

Typeset in Nimbus Roman
by KnowledgeWorks Global Ltd.

Contents

List of Figures

List of Tables

List of Abbreviations

HC	Healthcare
PHC	Primary Healthcare Center (PHC)
MIS	Management Information System
IS	Information System
DM	Data Mining
PM	Process Mining
ML	Machine Learning
BD	Big Data
DF	Data Files
LF	Log Files
EL	Events Log
BPM	Business Process Model
α Algorithm	Alpha Algorithm
β Algorithm	Beta Algorithm
ASD	Autism Spectrum Disorder
AI	Artificial Intelligent
TP	Treatment Procedures
IFF	If and Only If
ER	Emergency Room
SWOT	Strength, Weakness, Opportunities, Threats
E	Economic
R	Result
S	Scope
O	Objectives

Acknowledgments

As the first supporter of my research throughout the past few years, I acknowledge Prof. Sorin Ionescu, my doctorate's studies supervisor, for his continuous support. I would like to express my sincere gratitude for his patience, motivation, immense knowledge, and valuable guidance.

My sincere thanks also goes to Dr. Yehia Kotb and Dr. Seifaddine Kadry who provided me with an opportunity to benefit from their immense knowledge and research skills which added a lot to my journey. Without their precious support it would not be possible to conduct this work.

I would like to thank my friend and colleague, Dr. Soraia Oueida, who encouraged me to achieve this step of my life, and stands by my side in all situations. A special thanks to all who participated in the survey and who supported my work in order to get results of better quality. Finally, a special thanks to all hospitals general managers who showed appreciation and big support during my research.

Finally, I would like to dedicate this book achievement to my daughters, Yara and Lea, who I am sure will be very proud with my achievement when they grow up.

To my husband, Chadi, who supported me from the beginning until the end of this work to achieve my goal, to my all family members and especially my father who always inspires me to continue my research and improvements.

Thanks for all people who were part of this achievement, for your understanding and encouragement in all moments, especially the difficult times.

This book is just the start of the journey!

Preface

This book proposes a new framework for process learning, which is probabilistic, and it learns the process in an accumulative manner. The learning process is divided into phases.

First, by learning the set of events in many log files; log files can be events in a system or symptoms in a specific disease or disorder; and then calculating the dependencies among those events.

Calculating the probability distributions of all events comes after learning the dependencies and correlations between these events. However, the proposed solution and platform is presented in this book with many real examples that are related to healthcare.

The steps of prediction model and building the required knowledge are also presented in this book. The idea behind writing this book is to emphasize the correlations between the healthcare domain and technology in its different aspects such as managing records, information, and procedures; early detection of some diseases; and improving accuracy in choosing the right treatment procedures.

This book opens the way for healthcare providers to improve their services and increase patients' satisfaction.

Introduction

The industrial sectors have grown tremendously over the years, and traditional methods for maintaining business are not enough to face all the challenges or guarantee success for a long time; hence, the need to innovate new methods integrated with the existing ones is mandatory. Healthcare is a huge sector among both the industrial sectors and the service sectors and one of the most important domains. Improving this domain is a critical task because failures in healthcare systems have bad influences on many other sectors in life, and this impact was obvious on all life's aspects during the Covid-19 pandemic. In addition, healthcare systems own the biggest part of industry in terms of all the medical needs from medicines to equipment and any other health resources.

Improving the different healthcare system processes and managing them are complicated and costly tasks because of the huge amount and different types of information that healthcare systems generate every day. Moreover, healthcare systems are dynamic interconnected systems where different parties are participating to serve patients' needs and communities. The main focus of this book is to provide a new technique and method in terms of a new platform that is able to detect issues of the existing healthcare system which is implemented in any particular place such as a hospital or any other medical center, and provide solutions for those issues based on the extracted knowledge of the whole system. The probabilistic nature of this new technique helps to predict the high risk factors in any system and suggests solutions to show a huge impact on the system's performance. This new technique is deployed in the healthcare system to improve many important factors such as: decreasing the length of stay for patients in a hospital without increasing the cost or adding more resources; enhancing the treatment procedures by requesting the needed tests and neglecting unnecessary procedures; and reducing the deadlocks in the system, which helps to serve more patients, hence increasing satisfaction for both patients and managers. The advantages of process mining techniques to deal with processes and apply modifications on a system without dramatic loss of information about the current system model. In addition, process mining techniques can be integrated with other techniques such as management information system tools, analysis tools, big data, and, moreover, they are an intermediate phase between data mining and businesses' management process. Developing healthcare systems means to keep information safe from loss, detecting problems in the current systems without changing those systems dramatically or adding unreliable costs. All these goals can be achieved by applying methods which handle the current system and apply the required changes to improve it. The first phase in such methods is to access the log files which are files containing valuable information about all the activities in the system with many details. After analyzing those files, the knowledge about the system starts to be extracted to build the flow and model the processes of that system. Reaching that step provides a clear image about the current system with all its issues, challenges, and limitations; hence it allows the system analysts and specialists to find solutions for those issues. All process mining algorithms and techniques have the ability to extract knowledge, but they differ in how they can read the log files and what type of knowledge they can extract; where some of them are able to handle complex systems, others are limited to simple and small systems. The purpose of this book is to improve and manage the processes in healthcare systems using process mining techniques by studying all the famous existing different mining techniques, and innovating a new process mining approach applied on a real system in terms of a new platform designed for end users to detect all the possible knowledge aspects about that system, detect weaknesses, and find ways of solutions that can enhance the system with all its provided services. The improvement is related to the business model in the system and has a great impact on the services that are provided by this system which increases satisfaction for all parties in this domain. **The book structure is designed as follows:**

An introduction that presents an overview of the whole idea and the focus of this book.

- In Chapter 1, the healthcare system is defined from different perspectives and basic issues are discussed, and a case study about some managerial problems in hospitals is also presented. The chapter clarifies the need for implementing new methods, especially process mining methods, to overcome these problems.
- Then, in Chapter 2, a literature review about management of an information system, with its definitions and limitations, is presented in addition to reviewing different mining methods. After that, the IEEE data mining algorithms which are the most known algorithms are explained and their applications mentioned. A table of comparison between those algorithms is also presented in this chapter.
- In Chapter 3, the methods of mining processes are defined in detail, knowing that the focus is on the most famous methods in the world of mining processes. An overview of the Petri Nets and their correlations with the processes in the system is also mentioned. The α algorithm is chosen to be the base of the proposed solution in this book; hence it is explained in detail with all its facts and its limitations.
- In Chapter 4, a survey of basic process mining methods with their applications is presented; a new algorithm named the β algorithm which is an extension of the α algorithm is presented; the algorithms of how to detect events of log files, and the logical correlations between these events, their frequencies, and probability calculations steps are all presented in Appendix A. Also, in this chapter, the proposed solution is applied on big data of a healthcare system for a chain of hospitals and the output results about the system model are gained by integrating the β algorithm with Hadoop the big data engine. This step is important to show the robustness of the proposed solution toward the different scenarios and applications especially in the case of a huge amount of rapidly growing data in systems.
- A platform is proposed in order to provide the hospital or any related medical system with suitable, efficient, and user-friendly software, and is presented in

Chapter 5. All the work stated above is integrated in this new platform. The software helps managers and decision makers with proposing the best resource allocation for the smooth flow of patients and improving performance metrics during normal flow and critical events.

- The economic benefits and SWOT analysis are presented in Chapter 5, and Chapter 6 concludes the book and demonstrates the results with the outcomes of all the research.

The main focus of this book is using the concepts of process mining methods which are still new research topics and many of them are limited, with the aim of giving efficient results for real-life applications and especially for the healthcare industry.

Objectives and Overview

OBJECTIVES

The objective of this book is to improve healthcare systems in many aspects and highlight the issues in this important industrial domain by using the art of using data and process mining techniques to solve these issues. This book presents basic mining techniques and their methods such as data, text, image, and process mining in addition to big data concepts. Then it proposes a new process mining algorithm and framework called β algorithm that is implemented as "MYL" platform. This platform implements the concept of the β algorithm on real data files called the log files to detect knowledge about the current system and to suggest solutions for the limitations of the system business process model. This book presents a detailed explanation about the proposed β algorithm and the new proposed "MYL" platform with all the mathematical formulas and proofs; in addition, it presents applications that show the reliability of this new solution. The applications are tested using the proposed platform, and the output results are verified by many simulators such as Arena simulator. Moreover, the proposed platform is designed with the JAVA programming language which provides many features for ease of use for users who do not have a technical background. Moreover, the algorithm is tested using the Hadoop big data engine to manage a huge amount of data; hence, it makes it suitable for any industry or any domain which needs to manage big data for knowledge discovery and prediction. Finally, the β algorithm and its proposed "MYL" platform are tested from economic perspectives to show the robustness of the new solution by increasing the satisfaction of patients in hospitals and satisfaction of managers as well, by giving solutions that are able to reduce the cost and length of stay in hospitals from a patient perspective, in addition to improve the performance by finding solutions for many other business issues of the current system. The mentioned objectives with the activities and results of this book are presented in the following figure.

Objectives	Activities	Results

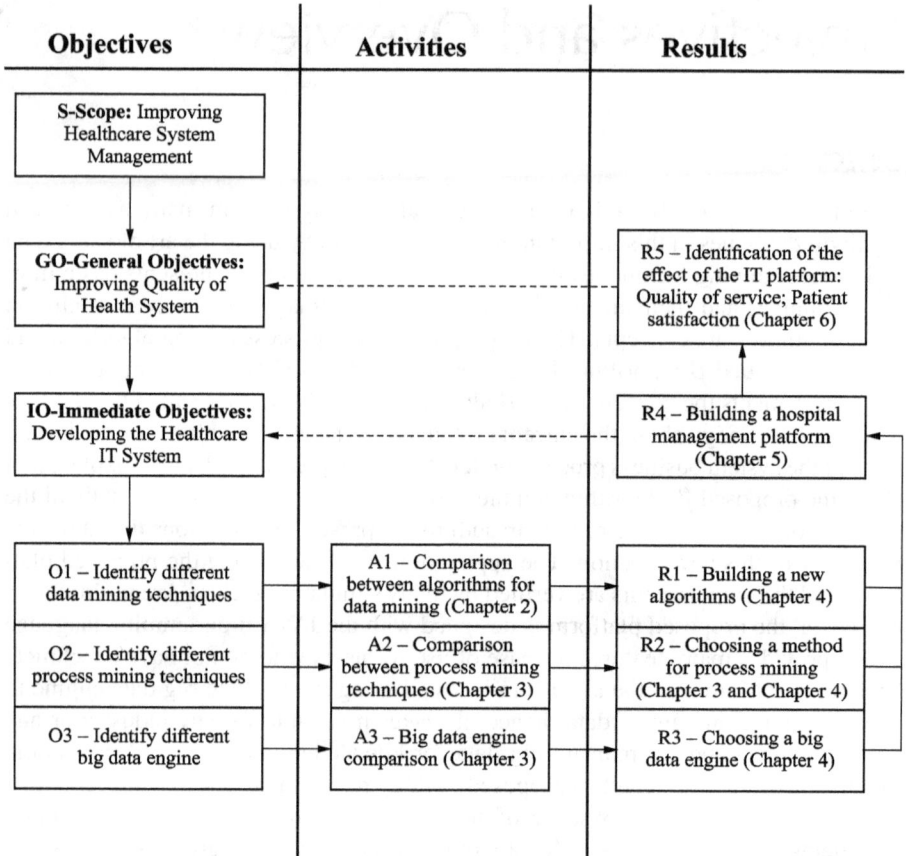

1 Healthcare System

INTRODUCTION

The high-growing business rates, the increased rates of charges in organizations, and the technological issues spur organizations to be effectively and continuously re-engineered and redesigned to achieve the success from strategic and operational perspectives. The main problem against achieving the desired result is the inefficiency of procedures and the lack of innovation which leads to weighty consequences for business and its competitiveness.

This chapter focuses on healthcare industry which is one of the biggest industries and service sectors in our world. Healthcare (HC) faces diverse challenges and difficulties since it is a very large area that serves various categories in the populace and has numerous parts in addition to different resources. However, HC system faces a lot of challenges because it deals with various aspects such as improving treatments, eliminating activities that have a poor effect, reducing expenses or time of waiting, serving a larger number of patients, and implementing new technological services. Furthermore, healthcare system has issues and challenges identified with operational management and inefficiently deployed processes.

This chapter defines the healthcare system and its segment with all the primary challenges that face the sector. It explains the reason for the selection of process mining techniques in this book so as to build up organizations and HC field specifically.

1.1 DEFINITION OF HEALTHCARE SYSTEM

Healthcare system (Hossain and Muhammad, 2016) is a system which uses the different techniques to develop and maintain the health of people that includes prevention, analysis, diagnosis, and treatment of ailments, diseases, or any sort of injuries, and even various physical or mental problems in persons. These procedures of healthcare system are offered in all medical sectors, such as hospitals, clinics, emergency rooms, where the services are rendered by health professionals in these fields. These health professionals include doctors and their assistants. Dentistry, midwifery, nursing, medicine, pharmacy, psychology, optometry, occupational therapy audiology, physical therapy, and other health professions are all part of the HC as well. HC includes services in provision of primary care, secondary care, and tertiary care, in addition to any related health services. Nonetheless, HC system can be defined in multiple ways, one definition describes and focuses on the goals that healthcare system seeks and another definition explains the elements that belong to HC system.

• Healthcare System According to the Goals

If the health systems are defined according to their goals, they serve as improvement for the health in general which is required as an improvement in the various aspects of

DOI: 10.1201/9781003366577-1

the health system such as its treatments, policies, techniques, and development of the economic benefits for the medical staff as well (Gilson 2003; Mackintosh, 2001). In addition, the goal of health system is not restricted to the definition above; it includes a broader range such as equity, the costs of financing the health system and fairness that means the balanced distribution of health, in addition to protect households from the catastrophic costs associated with diseases (World Health Organization, 2007).

• Healthcare System According to the Components

Healthcare system can also be defined according to its elements. It is a group of people, human and non-human resources who build healthcare organization to carry out the necessary health services to target populaces that need the services. Each country has its own wants in relation to the health services, and the HC system is programmed according to the needs of each country; therefore, there are variety of HC systems all over the world. Despite the fact that there are differences between HC systems all over the word, all health systems are categorized as primary and public healthcare systems. For example, the health system is distributed among market participants in some certain countries, in other countries, governments, trade unions, charities, religious organizations, or other co-ordinated bodies make greater efforts to make sure planned health care services are available to the populace they serve (Health system, 2017).

1.2   BASIC COMPONENTS IN HEALTHCARE SYSTEMS

Achievement of notable solutions to issues in the HC system needs a knowledge base of the different elements which belong to it. There are variety of elements that belong to the healthcare system. Examples are resources of the HC system, the provided services, and the departments.

The resources of HC system are categorized as:

1. **Human resources:**
    Human resources in HC system refers to every expert or professional who provide treatments to the populace. They include doctors, nurses, radiologists, technician, transporters, physicians' assistants, and all human that provide services for any health domain.

2. **Non-Human resources:**
    The non-human resourses can be described as all elements that serve patients in any HC area such as medical equipments, x-rays machines, ambulances, beds, wheelchairs, etc. All of resources above are known as assets of HC enterprises.

3. **Departments in Healthcare System:**
    Some of the departments in healthcare system are health providers such as hospitals, emergency departments, primary care, secondary care, pharmacies etc. This section provides a summary of these different departments.

• Emergency Department

Emergency departments (ED) are known to be complicated system which are faced by numerous challenges in management and accountability for most of many health-care (HC) expenses (Christensen et al., 2009; Walshe et al., 2010). In addition, the cost of treating patients is higher in the ED compared to other departments because some cases cannot wait and several resources have to be deployed for prevention. Furthermore, the need of giving quick actions for patients especially when they are not eligible to get the service that should be given to patients with urgent cases. Managing ED is a vital task for researchers to constitute progress in the health-care industry and maintain balance between the available human and non-human resources with the rapid services that are required continuously in the ED. However, some suggestions are availabe to provide efficient services and easy flow of patients without having bottlenecks or deadlocks, some of these suggestions are still tested to provide predictions about the needed services in the future, hence more accuracy can be achieved in reference to the required different resources, which affect the speed of patients' recovery and leaving the hospital, thus increasing the availability of those resources, to establish immediate response teams that act in very urgent situations in order to avoid mortality among others (Jensen et al., 2007).

Another suggestion is the "Fast-Tracking System," which provides an area for staff that is separated from the area of patients and it is a low acuity at triage for minimizing waiting periods, improving customer satisfaction, and reducing waiting time for getting served (Jensen et al., 2007; Leung et al., 2008).

Another method is named "Provided Directed Queuing" supplys development by providing a source which is "doctor at triage" to be one of the memebrs so that more accurate judgement and faster service can be provided, which creates a strategy to minimize the duration of waiting in the deparment of emergency services, especially when the ED is overloaded with duties. All the suggestions which are mentioned earlier are part of the ways of improving HC system in general.

• Primary Healthcare Centre

Another important entity in HC system is the Primary Healthcare Centre (PHC) This is an entity that discovers the general health problems and provides different health services such as preventive, curative, and recovery. It is a department saddled with the responsibility of providing service case for any issue or query. It cares over time, provides for all except exceptional or bizarre conditions, and coordinates care provided elsewhere by others. Additionally, PHC is an entity that provides the following features such as:

1. Being the front line in contact of various needs;
2. It focuses on persons more than focusing on diseases that need continuous care over time;
3. It is a total care for various and common existed needs in any society;
4. Coordinates care for common needs and the needs that rarely require special services (Lapão and Dussault, 2012; Starfield, 1998).

The major issue that faces PHC is the management of primary care compared with the analysis allocated to the management of hospital services in general. According to the studies, it is required to design health systems that have more focus on different types of medical care such as the "Primary", "Secondary", and "Tertiary" that are commonly found in hospital to support entities to primary care, and insisting on cases that are difficult to be served well in "primary care" settings (Walshe et al., 2010). Improving primary healthcare management is a great challenge because it is related to many aspects such as improving the "contracts, financial incentives", and the overall performance of the services in PH enterprises as well.

One solution with respect to HC management issues is to improve the management of the PHC, consequently increasing the effect of the PHC in the HC systems by improving the relations between its services and the hospital's ones. When this goal is accomplished, primary healthcare may become a source for developing health and improving stronger communities. There are lots of examples in this area such as U.K. healthcare systems where the general practitioner who is the medium to experts in the domain of "primary care" or directs patients to another sector such as the "secondary" or "tertiary" services while in some other HC systems such as Europe countries and even U.S.A., the patient can immediately refer to more expert physicians (Lapão and Dussault, 2012; Walshe et al., 2010).

• Pharmacy

Another department in the health sector which provides health services is the pharmacy department comprising of activities such as dispensing medicine, compounding and reviewing drugs' information or advices for efficiency and safety manners. Pharmacy is responsible for the guarantee of right usage of medical and healthcare products, thereby avoiding unmonitored self-medication. However, this department has lot of issues such as administration of the cost of services rendered, since this cost may appear to vary among different categories. For example, if a patient is hospitalized, the patient may receive free drugs, while in other cases may receive them as per the policy of the insurance companies of HC subsystem, or the HC system that patient is associated with it. However, a part of price will be charged in many situations taking into considerations some exceptions that can be applied on medicines for "severe" or "long-term" diseases. In Europe such as Portugal, the national health system pays more than the medium charge for what is called the "reimbursement of medicines" (Tribunal de Contas, 2011).

Cost management of pharmacies is a challenge because drugs companies like private sectors need revenue, while there have been adjustment made in the laws of some countries pointing to more difficulties related to drugs profit margins which has caused challenge in this sector because pharmacies are considered as high profit margins and without any competition because of the legislation in many countries. Recently, these have been in the rules and this is reducing the profit of those drugs, that increase difficulties for pharmacies when begin specifying their charges. *Antão* and *Grenha* have high expectations, which assume that minimizing the revenue if it is continued, hence many pharmacies may be close in the future, and as a result the number of unemployed people will be maximized in addition to the loss of tax's revenue (Antão and Grenha, 2012), which is also another problem encountered by HC systems that needs to improve policies and study the impact of new laws before implementing it on real scenarios or situations.

The following sections focus on HC system issues, their examples and suggestions for solutions in more details.

1.3 HEALTHCARE SYSTEM MAIN ISSUES

The healthcare system has a lot of challenges and issues. This section discusses these problems and the other sections in this chapter discuss the available methods to improve the HC system. The healthcare system is an important sector for societies and its problems can lead to the loss of patients, or death of people in mass daily from errors in certain situations, or accidents and infections in hospitals alone. Hence, these HC systems issues are weighty because of the catastrophe and havoc that may occur on a national level.

In order to solve and minimize the problems in any sector, the problems must be well defined and categorized according to their significance. Then, right solutions can be recommended for these issues. However, the main issue in HC systems is the hundreds of tasks that cannot be done initially. At the same time, the current systems are facing many technical issues that cause the employers and other purchasers who pay for the care of losing faith in having developments in the existed processes (Health system, 2017). Reengineering and redesigning the processes and procedures in HC systems can help to reduce some certain problems, however, some researchers believe that strong and reliable method to solve this main issue does not exist (Hoogervorst and Dietz, 2008). In Lifvergren et al. (2010) and Mintzberg et al. (1994), it is expected that over "70% of strategic initiatives such as Total Quality Management (TQM), Business Process Reengineering (BPR), and Six Sigma", among others, may lose because of the following reasons:

1. The loss of corporations between various objects in an enterprise at the modelling phase. Initially, many HC systems are deployed from long time.
2. The difficulties that occur by dealing with the dynamic changes in an enterprise especially at the operational level which comes from the weak initial models.
3. The loss of motivations to handle applications that can improve the "self-awareness" of any enterprise (Aveiro et al., 2010; Henriques et al., 2010).

All the above reasons motivate researchers to come up with more advanced and integrated ways of providing efficient techniques especially in this area. More about issues and challenges in the healthcare systems are explained in the next sections.

• The Gross Domestic Product (GDP) in Healthcare System

A dangerous but important challenge of healthcare system is its high demand for the Gross Domestic Product (GDP) in many countries, most especially in the developed countries. This demand is considered an increasing trend as shown in Figure 1.1.

This demand is posed as as such because of the lack of relationship between the service cost and quality inefficiency in resource consumption which affects the overall quality of the healthcare provided to population and affects the quality of life because of its result of the economy, tax rates and insurance contributions, disinvestment in other public services, and many problems to afford healthcare (Kaplan and Porter, 2011; Walshe et al., 2010).

Figure 1.1 illustrates how HC systems charges are more than "10%" of the "GDP" according to medium of various countries, the function of "expenditure" of the GDP increased to double in 50 years, with rapid growth. However, the largest one is recorded in the "United States, the Netherlands, and Portugal" (Walshe et al., 2010). This fact requires more further resources and effort to increase the GDP for the handling of the increased demand for healthcare systems.

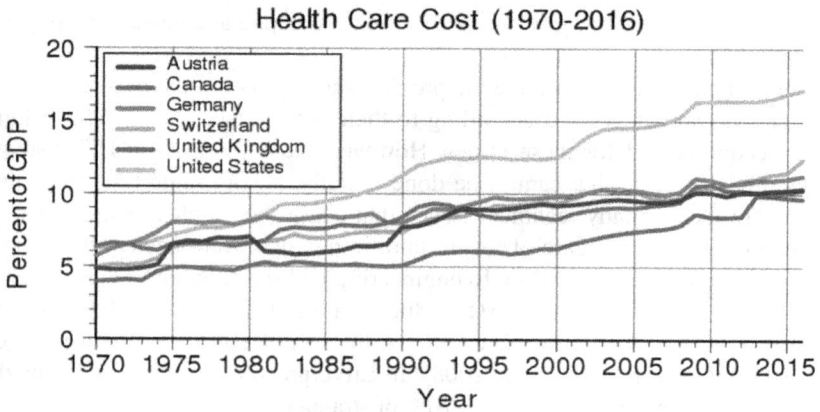

FIGURE 1.1
"Evolution of healthcare expenditure as GDP percentage", adapted from
Source: (OECD, 2012).

• Cost Crisis in Healthcare System

After unfolding the issues in HC systems that are related to procedures and processes in this domain and the increase in the demand on the total GDP of lot of countries. Another issue HC organizations face is the cost crisis that is caused by many factors including the higher demand of healthcare services, together with the rise in percentage of serious sickness, technology improvements, and the appearance of new treatments. Moreover, the policies of insurance companies and governments toward the costs of healthcare, face issues in the rising consumption of patients who do not care about the amount of charges they have due to the services they required (Kaplan and Porter, 2011; Walshe et al., 2010). The cost of healthcare services is a serious challenge, moreover, it is not easy to determine the required cost for higher quality healthcare systems because of the lack of accurate models or processes, such that, hence improving the healthcare sector is a challenge and a difficult decision to make according to (Kaplan and Porter, 2011).

However, the crisis of cost is also caused by lack of knowledge of the total cost required to provide care and services to patients which is required a systematic approach to the implementation of strategies that are reliable to various parties weather patients, service providers or who provide funds (Comission of the European Communities, 2007; Public Health Evaluation and Impact Assessment Consortium, 2011). The lack of balance between the cost and the required care leads to a reduction in purchasing power and a potential misallocation of resources. The suggested solutions are either denying the funding of the public services or incrementing of tax to extreme levels which makes it very difficult to provide HC systems with the needed services for elderly people or less privileged, which has a great impact on aspects such as the provided medical services (Christensen et al., 2009).

This can add competitions at the macroeconomic level through the diversion away the different parties from the productive ones. Moreover, it will have a negative affect especially on some countries that have huge expenses and will not be able to compete with others due to the unorganized variations between those countries in reference to their spending levels of "GDP" in HC systems as an example some are spending "17%" of their "GDP" while others is below "6%", as presented in Figure 1.2.

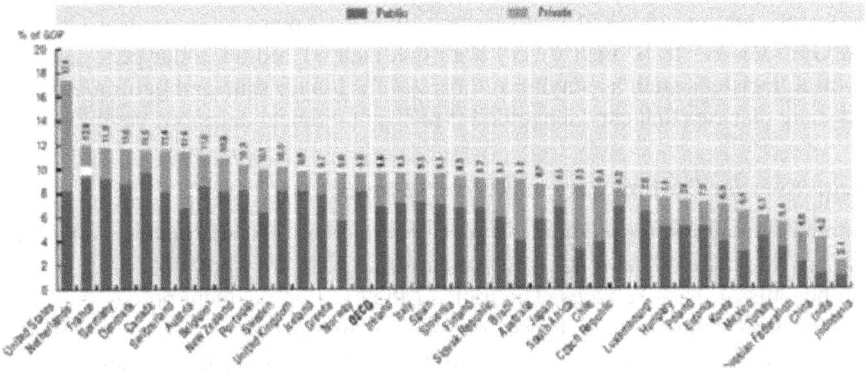

FIGURE 1.2
Healthcare expenditure as percentage of GDP in 2009, adapted from
Source: (OECD, 2012).

Whether the issues in healthcare systems are similar to technical issues, difficulties in procedure improvements, or the cost crisis, in addition to lot of other issues, the result of showing these problems, is that most of the processes of HC are directing their management to reduce its quality and efficiency levels. The focus in my book on how the process mining techniques are vital to overcome many situations that face organizations, especially the healthcare system organizations. The following section in this chapter suggests solutions and their value in solving problems.

The following section is a case study done by famous agency and presents how the wrong policies and procedures can cause damage to any organization dramatically.

1.4 STATE OF THE ART IN THE HEALTHCARE SYSTEM

The following subsections explain the great impact of some harmful decisions on healthcare system from nation's perspective.

• Too Much of Unnecessary Care

This issue is related to the unnecessary care services that are provided in many places and highest rate of healthcare expenses, and huge number of experts are losing productivity (Zhou, 2008).

• Non-Recommended Risks to Patients

The studies emphasize on some risky decisions that medical care beneficiaries take, such as early elective delivery where babies are born between 37–39 weeks which causes a higher danger of death for the babies, or increases the likelihood to have respiratory problems and afterward admitted to Intensive Care Unit (ICU). A study by the "Institute of Medicine Health" states that a big amount that is more than the third of health expenses are lavished, and as an example the unneeded "early elective deliveries" and are checked in a research of "The American Journal of Obstetrics and Gynecology" to be nearly very high that near $1 billion dollar per year. Other harmful decisions such as staying in hospital for longer time and getting infections and more diseases that can be avoided if the right medical decisions were taken.

• Expenses for Getting Care as Per-verse Incentives

In default situations, the medical insurance companies, or health planners, and Medicare provide the medical renders for whatever services they got, not considering the true benefits of those services. The issue of the existing payment processes in those systems is providing benefits for risky decisions such as the "elective early delivery". These processes can likewise accept admissions to ICUs since ICUs are centers to gain profit. However, researchers recommend that reduction of such risky decisions like these deliveries unless are needed can help to eradicate as many as one-half million ICU days, and reduce health costs on the country scale. Reducing the harm decisions such as the elective early delivery can cause reduction in profits for the hospitals but has a decent effect toward reducing the cost of medical services in the country. This should compel emergency hospitals to change their policies and procedures to keep away from such harmful decisions even if it reduces their profits but improves the overall performance of the health system because there will be focus on the real problems and diseases rather than spending money on unnecessary procedures.

• Lack of Transparency

Lack of information regarding the best choice for our healthcare is causing numerous startling procedures. Availability of transparent information about the consequences of any medical decision helps medical beneficiaries to avoid wrong and costly decisions.

With the knowledge of all the discussed issues, it can be concluded that the major problem and challenge that face HC systems is the inefficient processes that are implemented in the system, and that it constitutes a huge impact on the country and nations' health as well. Improvement in HC system can be implemented by encouraging a genuine change toward policies and strategies to develop and improve the systems without causing dramatic changes or loss of resources or data.

1.5   HEALTHCARE INDUSTRY MANAGEMENT SYSTEM

The motivation to improve HC systems is based on the importance of this domain as a sector that provides treatments to sick people and greatly affect all other domains and factors in any populace.

Although, the main issues in the healthcare domain are related to services efficiency, achieving successful strategic and operational policies, in addition to improvement of their business processes, but the improvement in HC domain goes beyond these goals by achieving improved treatments, minimizing unnecessary tasks, reducing the waiting time to get a service together with management of the expenses to avoid cost crisis for people and for governments or any other parties as well, treating more patients, and implementing new technological services (Christensen et al., 2009).

Knowing all these facts makes the aim of improving healthcare organizations more difficult and the improvements in the healthcare organizations are required not only because of the competition like any other industrial domains, moreover, is to maintain existence during the serious situations that may appear due to the shortage of technical improvements, the rapid growth in customers' demands, or other conditions that are related to either economy political views.

• Healthcare Structure in Industry

Healthcare systems services can be managed properly if they are classifying and analyzed as categories equals to three fundamental departments, which are the "primary, secondary, and tertiary" care, and modeled in Figure 1.3. Every health sector is analyzed and modeled as a subsystem of the whole industry. However, a patient who does not have to go to the emergency department starts communicating with "primary care" to be provided with startup services, and if those services are not enough for that patient, then the patient will be directed to the "secondary care" if there should arise an occurrence of serious sickness or needed advanced treatment. Afterward, a tertiary unit for extra follow-up. There is an overlap between these units, and it is often true that a patient may receive services from more than one sector at the same time (Lapão and Dussault, 2012; Walshe et al., 2010).

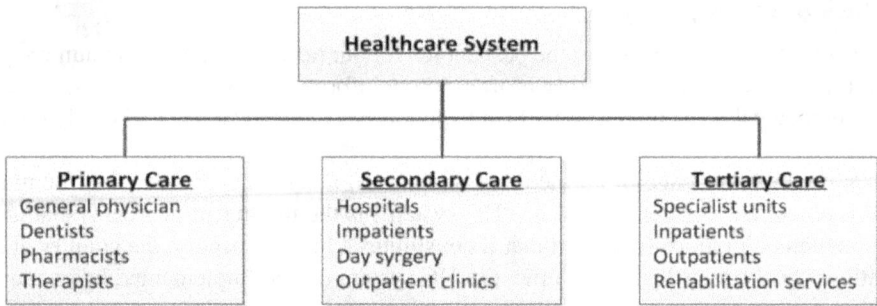

FIGURE 1.3
Healthcare system departments
Source: (Walshe et al., 2010).

HC management system is a complex and challenging system because of the overall economic expenditure that records from "8%" to "17%" in most powerful countries. However, improving HC management systems means to improve all the categorized sectors that belong to healthcare system in ways that have great impact on the various health factors that serve societies with rapidly growing rate and supply them with better services in this domain, in addition to studying new methods of treatment and diagnosis (Christensen et al., 2009; Jensen et al., 2007; Walshe et al., 2010).

• Business Models of Health System

One of the solutions to ensure improvement in HC organizations is to improve the healthcare business process models which affect the general productivity and performance. However, their models serve a big concern that can manage the human abilities to realize that process aspects, and implemented properly without breaking the stability among systems and their processes. However, there is no limitations on modeling processes for business aspects, and they can be modeled by using various techniques, tools, or even languages by only referring to standard "process modeling methods such as UML the Unified Modeling Language" (Rumbaugh and Jacobson, 1999), "ArchiMate" (The Open Group, 2009), "Petri Nets" (van Hee, 1994), "BPMN" (OMG, 2011), or even "ad-hoc" diagrams with some notion of activity. These models that are originally workflow display the series of activities to be identified. Then, a model design for a process displays the way of carrying out a business case (Caetano et al., 2011).

Improving a business process model is not an easy task in any domain because the modeling tools do not have all the techniques to analyze these models. It is predicted that above "70% of strategic initiatives such as TQM, BPR, and Six Sigma" among others, may not success according (Hoogervorst and Dietz, 2008; Lifvergren et al., 2010; Mintzberg et al., 1994). However, the lack of consistency and coherence between the different components of an organization is the main reason behind those failures. This fact gives the motivation of this book to study and analyze all the existed solutions related to different mining techniques and compare between them. Then, producing a new method and technique that overcomes many issues in other solutions and has a good impact on improving enterprises in the industry.

1.6 CONCLUSION

As observed and practiced, HC systems are mandatory for our societies, and they are considered important sectors in the industry because of the different sectors that are under the umbrella of healthcare system such as hospitals with all categories and departments, medical equipment's and medicines industry, health insurance services, medical staff, medical resources, and patients. All these components in the healthcare system need to be organized and managed by efficient methods that can keep track of all resources and provide continuous ways to ensure improvements and manage the medical staff and patients' satisfactions as well, in addition to managing the cost and required services. All the discussed requirements in healthcare and issues that challenge this service sector which some of them are stated in this chapter require strong techniques that provide improved model without causing the change dramatically, these techniques are data and process mining that are able to detect the current process in any hospital or health sector and highlights the advantages and disadvantages, not only that but also suggesting new ways of improvements. These features of using process mining techniques are explained in book especially in Chapter 3 all to solve the basic healthcare issues such as:

1. Providing a fully automated system for healthcare domains.
2. Reducing waiting time to serve patients.
3. Increase patients' satisfactions.
4. Improve the quality of services. In this book, the focus is on achieving the above points regarding healthcare systems by analyzing and studying different methods and come up with a new solution that achieves success in management information systems of hospitals mainly or any healthcare system, and overcomes many issues.

2 Data Mining and Management Information Systems

This chapter describes the management information system (MIS) in general, its methodologies, and the role of data mining in order to manage data in any enterprise. As seen in the previous chapter which shows that the healthcare systems organizations are faced with many issues and those issues are mainly related to the managing information in the system, and the procedures which are utilized in the systems. All of this information helps us to understand the meaning of MIS in any industry and how it affects various domains. Adding to MIS methods, mining techniques are also needed in any industrial domain to provide help for many challenges that the organizations could face, thus this chapter discusses the meaning of MIS and data mining with their popular techniques and their applications.

General management and management information systems (MIS) are the points to start discussing the meanings of data and process mining, where the management in general is defined as a process which is designed to ensure the cooperation, participation, intervention, and involvement of others in the effective achievement of a given or determined objective (Kassirer et al., 1991). It is related to management and the leading of an enterprise from the aspects below:

- The values and qualities of effective general managers and leaders.
- The methods that draw the strategies of the successful enterprises.
- The communications between any enterprise and other organizations.
- General management is divided into four categories which are
    - managing policies and processes
    - managing information systems
    - managing societies and enterprises and leadership
    - values, and corporate responsibilities.

2.1 BASIC ELEMENTS OF THE MANAGEMENT PROCESS

The management process consists of some basic elements, such as:

1. Decision-making: which is the key duty of all managers at all stages of the management process where they make choices among alternative courses of action.
2. Problem-solving: is more complex than decision-making when the choices among alternatives are made to overcome hurdles or limitations affecting the development toward the goal.

DOI: 10.1201/9781003366577-2

3. Human Relations: that can be implemented through leaders motivating others so that others will cooperate and participate among one other.
4. Communication: is the energizing force in an organization which governs collaboration and collective progress toward the goal. In summary, in every decision making process, the following are necessary for:

   (i)   Identifying problems and opportunities that require action,
   (ii) setting priorities.
   (iii) Determining the cause of all problems
   (iv) Analyzing the effects of problems.
   (v)   Defining how to choose sets of action to be performed.

• **Achieving Successful Process Through the Management Techniques**

The management concept consist of numerous applications, which includes the management science that identifies three main procedures, which are granting them the freedom to act within these boundaries, stimulating improved performance to define acceptable boundaries for actions of different levels of employees, and supervising them by results (D'Alessandro et al., 1999).

The following should be adopted to accomplish perfect successful procedure through the management concepts:

1. The realistic attitude.
2. Having a comprehensive system of rules of discipline. Keeping a sufficient system of communication.
3. Developing an objective follow-up pattern.
4. Maintaining a sound organizational atmosphere.

In conclusion, management is a dominant factor in bringing success to any organization and sustaining it in the market. Management comprises of many phases and one of the most important phase is to the availability of information systems capable of implementing better management decisions to build a successful process model. The need of good procedures in any organization came up between 1965 and 1975. It led to the automation of the functions of any organization. Payroll, stock controls, and invoices are examples of those functions, and they could only be found in big companies. After some years, technological advancement resulted in to what is called information systems (IS) to corporate policy, adding to the increase of the Internet in mid 1990s all affected the market and challenged to provide better approaches of doing business 121 (der Aalst, 2011a).

2.2   MANAGEMENT INFORMATION SYSTEMS (MIS)

• **MIS Classifications**

MIS can be classified according to many classes:

According to their formality, they can be human information systems which are named as informal information systems where sense organs are used by everyone to receive impulses from the environment. Another form of human information process is studying. Paper-based information systems is also a form of human information process where the papers were still being used because of their low cost and simplicity. The last one in this class is the computer-based information systems which are based on electronic methods to collect data, store, and process it and provide papers whenever they are required. Retailing, financial services, travel, and manufacturing are all examples of computer based systems (der Aalst, 2011a).

• According to Their Purposes

The purpose can be

Operational: it is the system of transaction processing and exchanging data between organizations.

Monitoring: it is the system of ascertaining qualities by monitoring the performance of the people, functions, services, and activities.

Decision support: it is the knowledge to managers that has been provided by the system in many different situations, mostly when they have to determine the consequences of specific actions.

Communication: it is the system that provides transfer of information among people and different organizations electronically like websites and emails.

• According to Their Reach

This class categorizes the information systems based on the location of their operations affecting their influence on organizations. These systems are individual systems, local or department systems, company-wide systems, inter-organizational systems, and community systems. The combination by different classes of the IS like purpose and reach gives some clues about the relative difficulties of implementing different systems (O'Brien et al., 2011).

• MIS Flows with Enterprise-wide Systems

The objective of enterprise systems is to set up an integrated platform for management between internal processes. The enterprise resource planning (ERP) is an example of the integration between the transactions-oriented data and business processes in any organization, ERP systems can be customized to suit any organization. Moreover, the business processes can be modified to fit with the system. Applying ERP system is only an organizational change process, rather than being a replacement of a technology. ERP systems has effect on people, culture, strategy, structure, decision making, and many other aspects of the company (der Aalst, 2011a).

• Knowledge Management System

KM is applied to improve how any organizations care, acquire, capture, store, share, and use knowledge. The KM systems depend on the data mining and different expert systems. KM is very helpful to any organization because of how it deals with organizational issues like the structural and the cultural (der Aalst, 2011a).

There are two views of the knowledge management process:

- Cognitive model and Community model.
- Include predefined objects and facts depending on the experience.
- Transferred through texts, and information systems.
- Transferred through social communication.
- The key factor is the human memory, and the key factor is the human community.
- The success factor is the technology, and the success factor is the trust.

• Managing Customer Processes with CRM

CRM is the customer relations management systems. It is designed to maintain the existing customers rather than attracting new customers. They are based on supporting and caring philosophy. CRM strategies are built on the fact that customers ought to be treated in some specific ways and depends also on adapting business process, structure and skills (der Aalst, 2011a).

2.3 MANAGEMENT INFORMATION SYSTEMS AND STRATEGIES

Some facts are necessary and they should be used with any information system in a business that successfully considers the right strategy that is best for that particular organization in order to enhance productivity and performance, these facts are:

- Improving the quality for customers.
- Reducing costs and works more efficiently.
- Differentiating products or services.
- Offering new or better products or services.
- Locking in suppliers or buyers.
- Raising barriers to market entrants.
- Improving employee's satisfactions.

• Information Systems from Strategic Perspectives

Information systems (IS) play a vital role in any organization, this role is of a different advantages from a certain point of view, such as increasing barriers to newcomers, and ability to dive into new markets, changing the balance of power in a profitable way, developing new products or services, reducing costs and implementing more effective management techniques. IS can show disadvantages by posing threat for same reasons discussed above where competitors always do the same (der Aalst, 2011a). There is no ideal information system for any organization, and different alternatives can be implemented depending on the business and its environment requirements.

Moreover, any IS can be influenced by the stability of the organization, the tasks of the organization, and the level of dependency among those tasks. The previous situation also affects the choice of the strategy to make it fit the business and customers' requirements, in addition to the way IS is used in organization which is to establish alignment between the strategy and the information system (der Aalst, 2011a).

2.4 INNOVATING BUSINESS PROCESS

Business process is defined as a collection of interrelated tasks, carried out to achieve a business outcome which is a chain of tasks from purchasing to manufacturing to selling and delivering. Business processes can be divided into two kinds: operational, which is related to the core business, and management, which include the IS and strategic decisions (O'Brien et al., 2011).

Process innovation is the implementation of a new method for production by developing new products or services, or changing the business model, or by developing new method of delivery. The innovation in any business can be achieved by changing the technology and environment of any origination. The advantages behind the process innovation like improving quality, faster processes, reduced labor costs, reduced materials, energy consumption, and environmental damage.

• Managing Process Innovation

In order to achieve the successful innovation process, the design must be able to manage different aspects of any project like IS, politics, people, interaction between different organizational aspects, and financial resources, moreover human situations should be given more attention due to the issues that rise because of the interaction between people and the change of the nature of their jobs (O'Brien et al., 2011).

Information systems play a major role in deriving the process change, and facilitating easy access to process information across functional boundaries. One of the issues is relating IS used to support the design to the overall IS strategy.

An example of the IS enabled process change is use of ERP which is implemented to integrate information resources and innovate process. ERP model must be selected very carefully because of its impact on the overall performance on any organization. Another example is deploying the business over the internet which leads to great change in the processes and their design (O'Brien et al., 2011).

• The Costs and Benefits of the IS

Implementing information systems requires cost, and in any investment, the idea of cost is related to the benefits which are gained after payment of the cost. The cost of any information system is divided into many aspects which includes: cost of purchase, hardware costs, software costs, and the cost of implementation and maintenance. However, the previously mentioned aspects are extended by the true cost of a new IS because many other things like the cost associated with the impact of the new system on the staff, customers, suppliers and other stakeholders should be taken in consideration. Moreover, it is difficult to predict the long term costs. The benefits of the IS can't be measured as other aspects of the business because some of these benefits could be tangible and can be detected by reducing costs of production or any other service required, staying in the market, improving quality, and increasing the revenue of the business while other benefits could be intangible making it very difficult to be measured like the staff improvement, customer management, value chain management, and flexibility in dealing with market changes (O'Brien et al., 2011).

However, nowadays, MIS methods are not sufficient to maintain any organization in the market and provide efficient solutions for all the main problems that could appear in any organization, for this reason, another technique called data and process mining which is related to data and strategies should be implemented along with MIS methods in order to achieve success and avoid dangerous situations. Hence, the following sections in this chapter presents mining techniques and their classifications.

2.5 THE STATE OF THE ART OF DATA MINING (DM) TECHNIQUES

The correlation between the management information systems and the mining techniques is the knowledge extracted from the data that is recorded by the information system applied in the organization (Weijters et al., 2006a). Terms such as big data, data mining, business intelligence, process mining, and other related terms, all have the meaning of analyzing and processing huge number of data, where this data is provided from the information systems (IS) as mentioned earlier. Nonetheless, all the mentioned terms have a common meaning, but they are different and the differences are discussed in the following sections.

There is a huge change from data orientation to process orientation. As I showed, "data refers to properties and features of the organization's element. For example, in hospital information systems, the data elements refer to the age, gender, diagnose etc. of patients while the process means the method of doing the tasks and giving orders of each task using the available resources of any organization (Weijters et al., 2006a). However, data and process mining have a lot in common. Both techniques are part of business intelligence, and the analysis of large volumes of data in order to achieve greater insights. They use same approach in their processes. Both data and process mining apply specific algorithms to data in order to uncover hidden patterns and relationships. The goal of data and process mining is to provide users with better decisions" (Zayoud et al., 2019a).

• Data Mining

It is data or knowledge discovery; it is the process of analyzing data from different perspectives and summarizing it into useful information that can be used to increase revenue, cut costs, or both. The challenge of using data mining is to find correlations or patterns among large data sets in order to convert these connections into information which is powerful for any organization performance as mentioned.

TABLE 2.1

Comparison between Different Mining Aspects and Technologies

Topic	Methods	Applications	Levels of analysis
Data mining	Consists of five major methods: 1. Extract, transform, and load transaction data onto the data warehouse system. 2. Store and manage the data in a multidimensional database system. 3. Provide data access to business analysts and information technology professionals. 4. Analyze the data by application platform. 5. Present the data in a useful format, such as a graph or table.	1. Used today by companies with a strong consumer focus – retail, financial, communication, and marketing organizations	• Artificial neural networks • Decision trees • Nearest neighbor method • Rule induction • Data visualization
Text mining	There are so many techniques developed to solve the problem of text mining that is nothing but the relevant information retrieval according to user's requirement, and the information retrieval basically there are four methods used: 1) Term Based Method (TBM). 2) Phrase Based Method (PBM). 3) Concept Based Method (CBM). 4) Pattern Taxonomy Method (PTM).	Emerging applications, such as: 1. Text understanding. 2. Electronic information is only available in the form of free natural-language documents rather than structured databases. 3. Search engine, text categorization, summarizing, newline, and topic detection	• Information retrieval • Exploratory analysis • Concept extraction • Summarization • Categorization • Sentiment analysis • Content management • Ontology management
Reality mining	It is using the big data to conduct research and analyze how people interact with technology every day to build systems that allow for positive change from the individual to the global community.	1. The usage of wireless devices such as mobile phones and GPS systems providing a more accurate picture of what people do. -Social networks.	• Detection algorithms • Community structure of the communication network • Graph analysis

(Continued)

TABLE 2.1

(*Continued*). Comparison between Different Mining Aspects and Technologies

Topic	Methods	Applications	Levels of analysis
Big data	1. Association rule learning. 2. Classification tree analysis. 3. Genetic algorithms. 4. Machine learning. 5. Regression analysis. 6. Sentiment analysis. 7. Social network analysis.	Numbers entered into cell phones, addresses entered into GPS devices, websites, online purchases, ATM transactions, and any other activity that leave a digital trail. Social networks.	• Process of examining large data sets containing a variety of data types. • There are many tools to analyze the big data that can improve the services provided.
Process mining	It is kind of technique that is needed in case of absence of valid description of a process or when the quality of existing documentation is questionable.	Monitor trails of management system functions, their logs of transactions that output from a resource planning system, or organizations, and products	• Specialized data-mining algorithms are applied to event log datasets in order to identify trends.

• Text Mining

The valuable information from various text data can be extracted, and this approach is referring to "Text Mining" and it is part of gaining knowledge as an output of databases of those texts. The hardest part is getting the right knowledge in those text containers to meet the required purposes.

• Reality Mining

It is the collection and analysis of machine-sensed environmental data pertaining to human social behavior, with the goal of identifying predictable patterns of behavior. Reality mining uses big data to conduct research and analyze the way people interact with technology every day to build systems that allow for positive change from the individual to the global community. Reality mining also deals with data exhaust.

• Big Data

It is the application of specialized techniques and technologies to process very large sets of data that are presented as huge and complicated sets that are difficult to process using on-hand database management tools.

• Process Mining

This type of mining is considered as a knowledge extraction, i.e., it focuses on log files and their events that are recorded by an information system applied in the organization (Weijters et al., 2006a). Despite the fact that the information systems help to get what is called log of events, these logs are not used to detect existed processes of an organization. Hence, the mining techniques aim to develop and discover a process by depending on data, that is, originally provided by those log files (Weijters et al., 2006a). It is a result of the need of emerging between data mining and business process management, where data mining focuses on large data sets and the business process management focuses on modeling these data. Hence, the process mining plays the middle-ware role between the two, to combine the data analysis with modeling. Moreover, process mining can handle the raw data or the event logs of any organization as proved in Zayoud et al. (2018a).

It can be concluded after considering the mentioned information in the comparison table between different mining methods that each type of mining technique has its own features and can be used for specific task, and when it comes to healthcare systems, all the mentioned mining methods are needed to build a sophisticated system starting from the phase of collecting and organization data and then mining data to give valid and useful information, then mining the different types of data such as texts, images such as x-rays, or any other type of data related to patients or the medical domain. Then building and managing data to perform reliable procedures required another mining technique which is named the process mining that builds processes and improves existed processes by monitoring the system model and its strengths and weaknesses in addition to keep track of any problems that may appear and affect the system later. Big data is also one of the mining techniques and is needed when the system is growing rapidly and required different types of procedures to handle huge amount of information appear in the system.

2.6 DATA MINING VS. PROCESS MINING

Data mining is a method that detects, analyses, and discovers data, it has no relation to the business processes directly while the process mining discovers, controls, and improves the actual business processes based on the gained data from information system implemented in an industry. By analyzing data derived from the IT systems that support the processes, process mining gives a true, end-to-end view of how business processes operate. Data mining analyses static information which is the available data at the time of analysis. On the other hand, process mining looks at how the data was created. Process mining techniques also allow users to generate processes dynamically based on the most recent data. Process mining can even provide a real-time view of business processes through a live feed which is I wrote in the article (Zayoud et al., 2018a).

Data mining searches for anything strange in the collected data, without considering answers to the question that might come up regarding this data, while the process mining gives room for answers and clarification of predefined questions.

TABLE 2.2

Comparison Table between Data Mining and Process Mining

Method	Business process direct link	Business process opera-tion	Static infor-mation	Dynamic processes	Concerns with hidden data	Results manipu-lations
Data Mining	✗	✗	✓	✗	✓	Limited
Process mining	✓	End-to-end view	✓	✗	✗	✓

2.7 DATA MINING IEEE ALGORITHMS

Defining the mining methods of data and processes are found in this section, their algorithms and techniques, and the differences between them. As seen earlier in this chapter, the techniques of data mining are implemented in various domains to give meaning to the available data. This section identifies without losing the generality the most famous algorithms and they are widely implemented in various mining domain in terms of data and based on the "IEEE International Conference on Data Mining (ICDM, http://www.cs.uvm.edu/~icdm/)" those algorithms are recognized as the most ten algorithms of mining in reference to data part. "This section explains the top 10 algorithms of data mining, such that Apriori, C4.5, PageRank, k-Means, SVM, EM, Naïve Bayes, AdaBoost, kNN, and CART" (Sun et al., 2011), the next section is presenting some of the most recent data mining algorithms like U-Apriori, UF-growth, UFP-growth, UH-mine, PUF-growth, and TPC-growth.

• C4.5 Algorithm

"Another famous DM algorithm is the C4.5, which is set of algorithms used for clas-sifying problems in DM or ML (Quinlan, 1993). It is necessary if there is a need to learn mapping between attribute values to classes knowing that these classes are used to classify new, unseen instances. Defining instances can be shown as groups of characteristics that classified as an object of various mutual exclusive objects. Tables that are built based on this algorithm shows the instances as rows, rows map to spe-cific values, and attributes that mean different meaning such as forecast information, while the "class" means whether the conditions are conducive to some rules such as playing soccer, for example. All information in the table is described by values for attributes such as temperature, humidity, and all of data in the table constitutes that may named "training data", as an example, hence the goal is gaining information to achieve proper predicted values regarding the arbitrary variables of a particular class. We can say that "C4.5" is ideally used in applications that are not considered as "de-cision trees", which are not similar to the sequences of help questions that clarify the root of any issue. C4.5 is to restate its trees in comprehensible rule form. In addition to inducing trees, the rule post pruning operations" (Ramakrishnan, 2009a).

Since C4.5 algorithm is a decision tree learning algorithm, hence it creates a flowchart to classify new data. It can be applied in health domain to help doctors to take decisions based on some information that builds the decision tree. For example, a doctor can decide treatment based on information related to a patient such as,

- Previous historical inforamtion.
- Existing genes that strongly related to cancer.
- Tumors detection.
- Sizes of tumors in that patient exceeds 5cm.

The above information is building a flowchart with the value of some attributes, and depending on those values, the doctor can decide a path of treatment.

• k-means Algorithm

The k-means algorithm is one of the most used clustering algorithm. "clustering algorithm is defined as procedure that divides given entities into categories or "clusters" where those entities in a category should be similar in comparison to others that belong to other different category. In more details, the algorithms that are clusters based set same data within one category while the different ones are set in other category. The main issue of classification or clustering techniques that their contents do not have a specific goal and those approaches are usually match the applications that have difficulties in categorizing their data. In this situation the categories are established based on characteristics of existed entities" (Ghosh and Liu, 2009). Those characteristics are recognized based on application nature (Wu and Kumar, 2009).

The above resultant clusters are identified as "poor" because they are not match perfectly with the "True", implicit portions.

"As a conclusion: The k-means algorithm is a simple iterative clustering which has an aim to separate a dataset into k clusters.This algorithm is based on iterating over two steps: (1) clustering all points in the dataset knowing the distance between each point and its closest cluster representative and (2) reestimating the cluster representatives. However, the limitations of the k-means algorithm are the sensitivity of k-means to initialization and determining the value of k. Despite its drawbacks, k-means remains the most widely used partitional clustering algorithm in practice. In summary, the algorithm is simple, reasonably scalable, easily understandable, and can be easily modified to deal with different scenarios such as semi supervised learning or streaming data. The continual improvements and generalizations of the basic algorithm have ensured its continued relevance and the increased of its effectiveness as well" (Ghosh and Strehl, 2006).

As an application of this algorithm, we can assume that we have a dataset of patients in a hospital. This information is called observations in reference to cluster analysis. Each patient's information related to the weight, height, pulse, age, blood pressure, choleserol, VO2max, etc. is a vector representing a patient. On other hand,

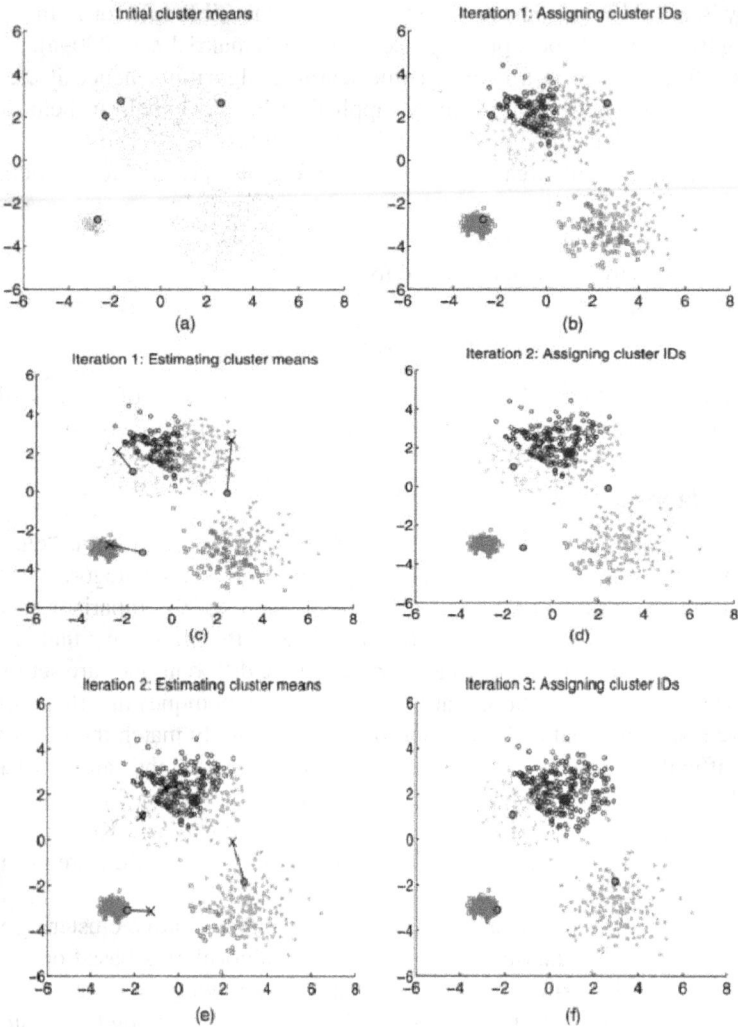

FIGURE 2.1
k-means on artificial data
Source: Van der Aalst et al. (2004a).

we can generate another vector that is related to the numbers of that patient, that is a date list to interpret coordinates in multi-dimensional space. Knowing that pulse is only a unique dimension, blood pressure is another one, so on. Having those sets of vectors, helps to cluster patients together if they have equal values of pulse, age, blood pressure, etc. "K-means" supplys the count of clusters, and picks marks in multi-dimensional space to express every *k* cluster. These are named centroids.

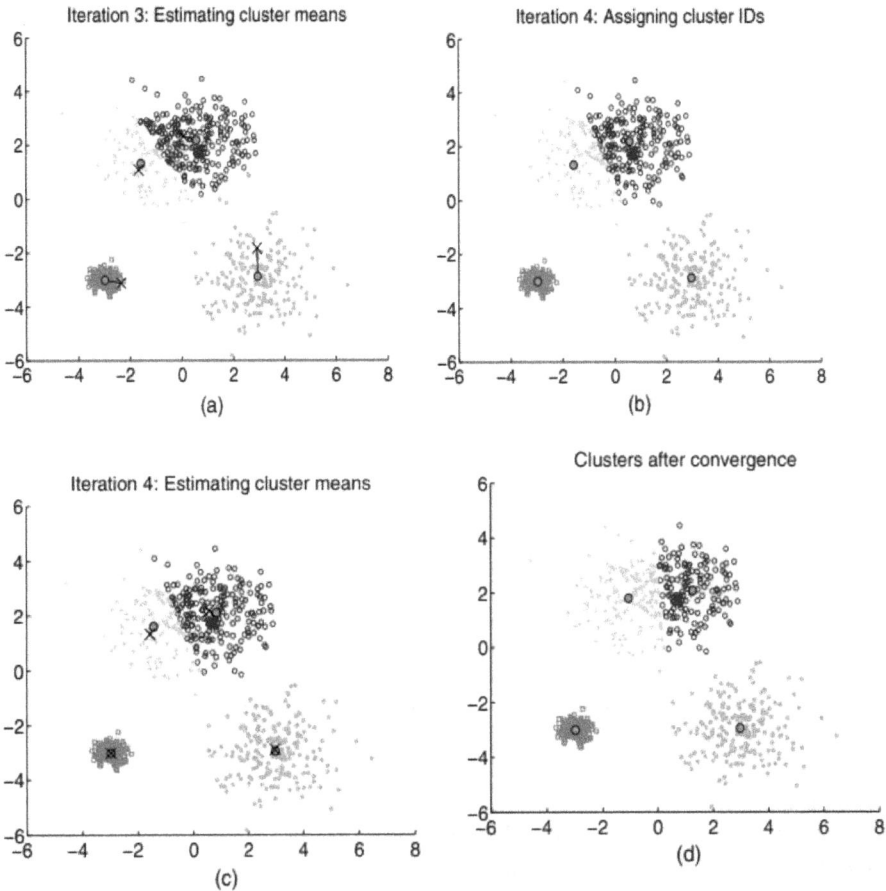

FIGURE 2.2
(Continued from Figure 2.1 k-means on artificial data.)
Source: Van der Aalst et al. (2004a).

- Every patient may be close to a part of k centroid; but not suppose to be for the same one, otherwise all have a cluster around their the nearest centroid.
- What we have are k clusters, and each patient is now a member of a cluster.
- k-means then finds the center for each of the k clusters based on its cluster members (yep, using the patient vectors!).
- This center becomes the new centroid for the cluster.
- They possibly exist around other closer centroids and choose another cluster membership.
- When the centroids are not possible to be changed, hence the steps "2–6" are stopped from repetition, and a stable memberships of clusters is achieved, and this is called convergence.

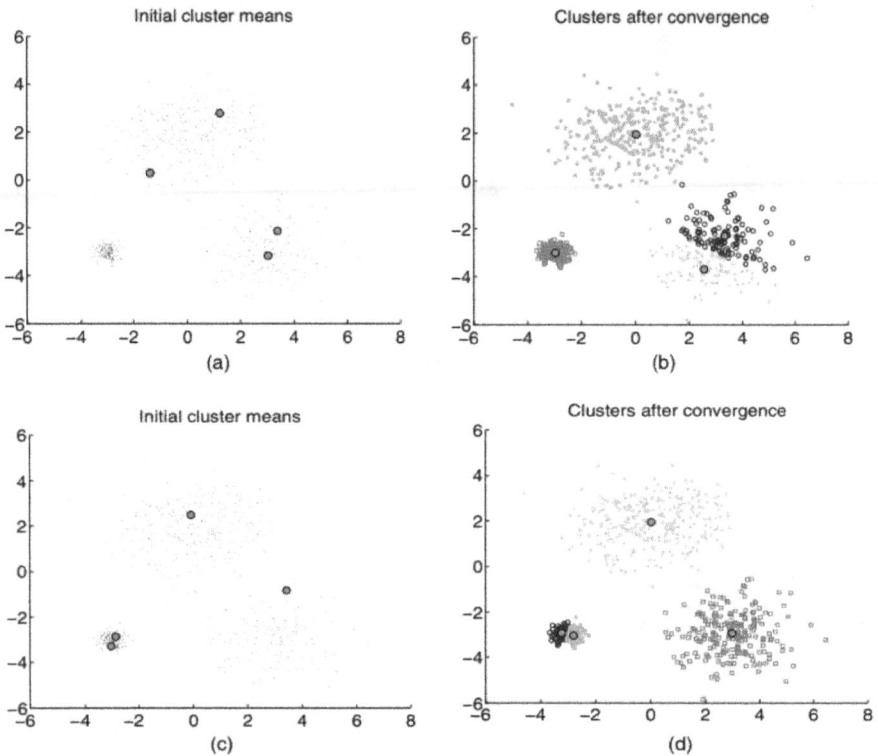

FIGURE 2.3
Samples about impact of enrich declarations on clusters
Source: Van der Aalst et al. (2004a).

• SVM: Support Vector Machines Algotithm

"Another interesting method is named Support Vector Machines (SVMs), which also includes Support Vector Classifier (SVC) and Support Vector Regressor (SVR), those are considered the most robust and accurate methods in the well-known data mining algorithms. However, SVMs is originally developed by Vapnik in the 1990s (Vapnik and Vapnik, 1998; Vapnik, 1995), it belongs to statistical learning theory, and requires considerable number of examples that can be used for training, and are often insensitive to the number of dimensions. SVMs have been developed at a fast pace both in theory and practice. Moreover, SVC aims to find a hyperplane that can separate two classes of given samples with a maximal margin, which has been proved its ability to show the best generalization that refers to the fact that a classifier is not only has good classification performance or accuracy on the training data, but also guarantees high predictive accuracy for the future data from the same distribution as the training data" (Fung and Stoeckel, 2007).

SVM: Support Vector Machines

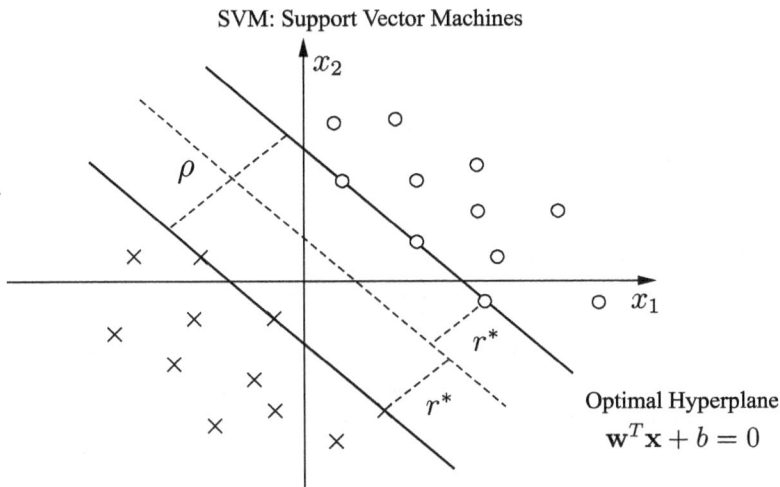

FIGURE 2.4
Illustration of the optimal hyperplane in SVC for a linearly separable case
Source: Van der Aalst et al. (2004a).

• Apriori Algorithm

"The decision tree building which are considered as pattern finding algorithms which are similar to classification rule induction, and data clustering that are frequently used in data mining have been developed in the machine learning research community. The association rule mining and the frequent pattern is one of the few exceptions to this tradition. Its introduction boosted data mining research and its impact is tremendous" (Sun et al., 2011). The main algorithms are considered simple and easy to implement, and the approach apriori is a simple and basic approach about many aspects such as its weaknesses in its databases to point transaction, dealing with limited data types, or even having inefficient calculations. However, there are continues development to overcome those weaknesses (Wu and Kumar, 2009).

• EM Algorithm

The "Expectation-Maximization (EM)" method, that is, used widely to substitute the other approache of calculations which is the "Likelihood ML Estimates". Moreover, it handles the limitation in data when it is in a noisy file. It is practical because of handling the issue of matching "finite mixture models by ML", that are deployed to design nonhomogeneous models of managing clusters or identifying contexts of pattens. This algorithm contains many stable aspects such as the numerical aspects, or the "reliable global convergence". However, many improvements are applied on this algorithm to solve complicated situations that are categorized as mining problems. The main feature of it, it is straight forward and has many stable sides (Wu and Kumar, 2009).

• PageRank

"PageRank has great impact on the web because of its powerful tool to analyze models. However, this approach is established to solve various complications of the "Ranking" algorithms that are related to their contents for the search purposes that have concerns about extracting only pages with the matched content to the requirement of a search query, but after 1990s, and due to the rapid growth of the web pages contents those approaches that depend on the matching of contents in getting required pages are no longer valid. To clarify the mentioned information we can give example about how "Google search engine" is providing references for a search query of the sentence "classification technique", in this scenario of PageRank should expect almost ten million relevant pages, and in this case applying the concept of scoring and ranking is problematic because of the huge available similar information in those millions of pages while having only few pages that are fit the query helps to rank them accurately for the user. Moreover, some owners of web pages reuse some of the known keywords frequently to give their pages higher score and rank whenever there is a search query even the content of their pages is irrelevant to the meaning of that query", (Chen et al., 2002).

• AdaBoost

AdaBoost, short for "Adaptive Boosting", is a machine learning meta-algorithm formulated by Yoav Freund and Robert Schapire who won the Gödel Prize in 2003 for their work. It can be used in conjunction with many other types of learning algorithms to improve their performance. Researchers have devoted tremendous efforts to the pursuit of techniques that could lead to a learning system with strong generalization ability. One of the most successful paradigms is ensemble learning (Zhou, 2008).

• kNN: k-Nearest Neighbors Algorithm

"The Rote Classifier is one of the simplest and rather trivial classifiers, which memorizes the entire training data and performs classification only if the attributes of the examined object perfectly match the attributes of one of the training objects. An obvious problem with this approach is that many test records will not be classified since they do not match any of the training records perfectly. Another problem arises when two or more training records have the same attributes but different class labels. A more sophisticated approach, k-nearest neighbor (kNN) classification (Fix and Hodges Jr 1951, 1952; Tan et al. 2006), finds a group of k objects in the training set that are closest to the test object, and bases the assignment of a label on the predominance of a particular class in this neighborhood. This addresses the issue that, in many data sets, it is unlikely that one object will perfectly match another, as well as the fact that conflicting record about the class of an object can be supplied by the closest objects to it. There are several key elements of this approach: (i) the set of labeled objects to be used for evaluating a test object's class, 1 (ii) a distance or similarity metric that can be used to compute the closeness of objects, (iii) the value of

k, the number of nearest neighbors, and (iv) the method used to determine the class of the target object based on the classes and distances of the k nearest neighbors. In its simplest form, kNN can involve assigning an object of the class of its nearest neighbor or of most of its nearest neighbors, but a variety of enhancements are feasible and are discussed below. More generally, kNN is a special case of instance-based learning" (Steinbach and Tan, 2009).

• Naïve Bayes

"Provided with a set of items that is considered as an object with defined "vector of variables", that has a purpose of developing a criteria of distributing entities to different categories, but this vector has some issues that are known as "supervised classification". For this manner, a powerful technique is named "Naïve Bayes" is used for multi-purposes such as construction without the need of any complex iterative parameter estimation schemes. This means it may be readily applied to large sets of data. It is also easy to interpret; hence users with limited skills in classifier technology can understand why it is making the classification that it can produce. However, it is not the most optimal solution for categorization and classifications manners, but it is reliable and trusted to be implemented. For example, in an early classic study comparing supervised classification methods, Titterington et al. (1981) discovered that the independence model produced the best overall result, while Mani et al. (1997) achieved the result that presents the effectiveness of that achieved model to extract efficient prediction about "Breast Cancer Recurrence". More examples are also presenting the efficiency of the "Naïve Bayes" approach, all those examples are explained in Hand and Yu (2001). It can be said that the "Naïve Bayes" method is simple due to it allows to establish initially two or more categories with labels as "$i = 0, 1$", to achieve scoring criterion which depend on set of classes initially, and it has rules for the those scores where the class "1" has to be attached to the maximum score, and class "0" is given for the smallest score. Other classes are established by checking their score with a "classification threshold". When a high score that exceeds the "threshold" is achieved then their classes will be attached to the class "1", and to class "0" when their scores are less than the "threshold". The "Naïve Bayes" is described from other perspectives in the below representation" (Turhan and Bener, 2009).

The above table shows the different and most famous data mining algorithms, their features, their applications in addition to their limitations. The overview of those algorithms shows the need for more algorithms to handle the different needs for organizations and especially healthcare organizations that have increasing demands. However, data mining algorithms are useful in some phases in managing data of the different systems but also not enough to handle all other phase such as building processes and designing business models, hence the need of other types of mining techniques is a must and innovating new methods always required to handle the issues and problems in the systems. The next section explains other type of mining which is process mining and show its perspectives and different algorithms and how can be used in the different systems and especially in healthcare systems.

TABLE 2.3

Data Mining Algorithms (Survey)

Algorithm	Methods	Applications	Applications examples	Similarities
C4.5	Tree-based algorithms attests	1. Domains with nominal valued. or 2. Categorical various data types.	1. "Clinical decision making", 2. "Manufacturing, document analysis", 3. "Bioinformatics, spatial data". 4. "Modeling geographic information systems".	Many similarities appear between C4.5 and CART.
k-means	It is a clustering algorithm.	It is for classes that have points descriptions "d-dimensional vector space". Thus, it clusters a set of d-dimensional vectors.	Semi-supervised learning. Streaming data.	One of the clustering algorithm.
SVM	Belong to statistical learning theory.	It uses technique to solve linearly inseparable problems.	SVMs are helpful in text and hypertext categorization applied in the biological and other sciences. Hand-written characters can be recognized using SVM. Classification of images can also be performed using SVMs.	The clustering algorithm which provides an improvement to the support vector machines is called support vector clustering.
Apriori	Is an algorithm for frequent item set mining and association rule learning over transactional databases "level-wise complete search (breadth first search)" algorithm.	Apriori is designed to operate on databases containing transactions ().	Collections of items bought by customers, or details of a website frequentation.	Agrawal and Srikant are extended algorithms for Apriori. Apriori family (Apriori, AprioriTid, AprioriAll).

(Continued)

TABLE 2.3

(*Continued*). **Data Mining Algorithms (Survey)**

Algorithm	Methods	Applications	Applications examples	Similarities
EM	Distribution-based clustering algorithm.	Applied to design heterogeneous outputs for identifying patterns and objects contexts "iterative computation of maximum likelihood (ML) estimates", applied for different noisy-data problems, used for data clustering in machine learning and computer vision.	Ideally used in medical image reconstruction, especially in positron emission tomography and single photon emission computed tomography.	—
PageRank	The best known link-based ranking algorithm.	It is deployed as a fundemental approach to study and design web models, to rank websites in their search engine results.	Powers the Google search engine.	Mathematical algorithm based on the webgraph.
AdaBoost	Is a machine learning meta-algorithm.	It creates a hypothesis sequentially in addition to their weights, which can be regarded as an additive weighted combination.	Face detection.	Boosting is actually a family of algorithms.
kNN: *k*-Nearest Neighbors	Gives a summerized form of the large scale of closest-neighbor categorization method, a non-parametric method.	Classifier is a simple classifier that works well on basic recognition problems.	An example of identifying faces using "*k*-NN" characteristics of dimension reduction or processing steps.	Scientists achieved a fact that says that "*k*-NN" outperformed a "Support Vector Machine (SVM)" approach.
Naïve Bayes	A probabilistic approach and easy to construct.	Is related to probabilistic classifiers applications by applying Bayes' theorem using the power of (Naïve) independence assumptions between different features.	It is suitable for problems of classification such as identifying the gender based on some features.	Naïve Bayesian model is suitable for huge data sets similar to many concepts that manage large occurrence of data.

2.8 PROCESS MINING TECHNIQUES

Information systems (ISs) have been rapidly growing in addition to the increased demand in all kinds and sizes of companies on those ISs. Moreover, new systems are shifting from supporting single functionalities towards a business processes orientation (Burattin, 2015). In Engineering and Science fields, an innovative area of study is begun to emerge and its called Process Mining. Process Mining is rich of methods and approaches that are specialized for analyzing and improving processes in any system. The following sections present detailed explanation of mining techniques in terms of processes and their methods.

• Business Process Modeling

The needed actions of companies to complete their own business are not simple and depends on interference of several humans and non homogeneous systems. Decomposing operations tasks in any business into pieces has better management results" (Burattin et al., 2013). The term "business process" is defined as a group of different activities and interactions, as appeared in Figure 2.5 as a simple process in store which has some tasks as events representations. It is showing the process as a workflow or as a graph of events' dependency, each action is shown in an individual rectangle shape, and links between them provide the order of those actions, the process is according to the figure begins at the registering an order phase, and ends with the "Shipping" phase. Knowing that some activities are executing simultaneously and being independent from each other. Then the phases of Goods wrapping and 'Shipping note preparation', the last phase of the order process is "Shipping".

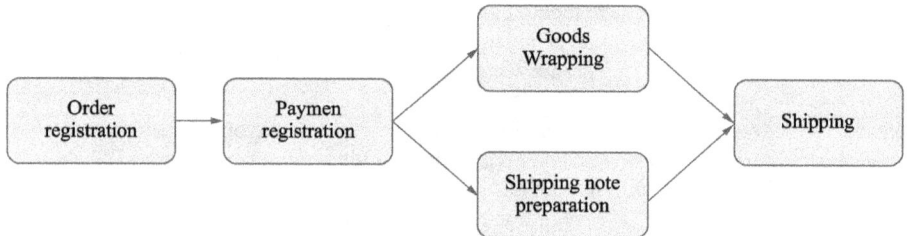

FIGURE 2.5
General process of order management

"Table 2.4 is another representation of the above process but in terms of traces that are actually the events of a process. As shown in the table, more details such as timings of events are listed in that table, where the time is playing a strong role to gain knowledge about any system's process and mine this process deeply to achieve the required goals. The importance of timing information affects various aspects of mining a process such as differentiating the event from any other general task not considered as an event and identifying correlations between those events in a log file. Moreover, time is a key factor in small and medium companies which do not perform their work according to a formal and explicit business process; instead, they execute their activities with respect to an implicit sorting that requires the time information. As a basic idea about mining approaches, having a log of events even with no clear model for current process as in Figure 2.5 is enough to gain a business process model consistence with the available recorded events. However, surfing into the domain of process mining in this book will provide the reader with different problems, and challenges that face achieving a business process model easily by only knowing events with their initial ordering as in the Figure 2.5" (Burattin et al., 2013).

TABLE 2.4

Process Representation of Two Implementations in the Business Process Described in Figure 2.5 – Source (Burattin et al., 2013)

No.	Task	Time Information	
First Log			
1	Phase of registering order	March 19,2016	12:30
2	Starting of payment	March 21,2016	09:00
3	Preparing products	March 25,2016	08:30
4	Special notes about shipping	March 25,2016	09:10
5	Phase of shipping	March 25,2016	10:20
Instance 2			
1	Phase of registering order	March 22,2016	15:45
2	Starting of payment	March 24,2016	17:30
3	Special notes about shipping	March 25,2016	08:00
4	Goods wrapping	March 25,2016	10:00
5	Phase of shipping	March 25,2016	12:30

2.9 CONCLUSION

As seen in this chapter, the different mining methods help organization to gain knowledge which is an indispensable resource. Knowledge management resources turn out to be a necessity for development. Only the useful knowledge has the significant approach for management and decision making therefore discovering this useful knowledge is essential. This chapter presents how data mining is a major part of knowledge management, and the classification of data mining tasks is the employed model in industrial domains for description and prediction. The implementation of MIS methods along with the suitable mining technique solves the different management problems. According to Table 2.3, there are many algorithms and approaches where each one of them is suitable for specific applications in industries, the table shows briefly the criteria of getting the best method for an application but the table did not recommend one methods over the others because each one of them has its own advantages and disadvantages which make it suitable for one application in industry but not for others. However, in the healthcare industry where machine learning and huge datasets are the main attributes of this domain, algorithms such as AdaBoost, and the Naïve Bayes can be the most suitable ones to mine data in the healthcare domain.

However, it is shown in this chapter that data mining can be integrated into the MIS framework to enhance the management process with better knowledge. It is understood that the data mining techniques will have a major impact on the practice of discovering knowledge and will present significant challenges for future knowledge and information systems research. In summary after studying MIS and basic data mining techniques and checking their methodologies, the next chapter will discover more mining techniques that are related to processes in organizations and emphasizing on choosing the suitable process mining algorithm to be the base of this book work to overcome the different issues and challenges that face organizations, and give those organizations the ability to evaluate their information systems based on the factors that are affected by the information system (IS) (Zayoud et al., 2018a); (Zayoud et al., 2017b).

3 Process Mining and Big Data in Healthcare

3.1 INTRODUCTION

This chapter focuses on the idea of knowledge extraction that can be used to improve procedures and processes of business models. The process mining is defined as the model's extraction for any business processes, and this model extracted from companies' information system in log files format which helps the internal and external activities of business organization to be understand and optimized. Those business models are presented in different formats such as the graphical to be used by analysts, or management, and the formal representation that can be used in mathematical analysis. However, process mining is an attractive domain for business development manner, hence a variety of algorithms and methods are being created (van der Aalst et al., 2009a). Those different methods and algorithms are increasing the need to select a suitable method for an organization. The right choice is required by both, the businesses using process mining, and the researchers evaluating new developments. "Moreover, an accepted process mining algorithm must analyze the log, then identifies all the process instances, after that it defines some relations among activities. When all the relations are available, it is should combine them to construct the mined model. For this purpose, many process mining algorithms have been designed and implemented, using different approaches and starting from different assumptions. However, even if several methods are available, many important problems are still unresolved" (Burattin et al., 2013). Some of them are presented in Van der Aalst and Weijters (2004b), and here the most important ones:

- The wrong connections between activities in the resultant model because in some models, the same activity appears several times, and each time in different position, which is difficult to be recognized by most of the available mining methods and this is considered a real weakness of this domain.
- The accuracy of the mined models is affected because the mining algorithms do not use the details of the available data such as detailed timing information that distinguishes the starting from the finishing time of an event.
- Current mining algorithms are not handling the varied perspectives, coming from different sources, hence they are not able to give more insights to create a global vision about the organization.
- The noise and incompleteness are two main issues that affect results of the mining algorithms that have limitations to handle those two issues, where the noise means the unidentified behavior of some activities in the log file, and the incompleteness means the lack of information that are required for performing accurate results in the mining tasks.

DOI: 10.1201/9781003366577-3

• The results' presentation of process mining can be as visual format to improve the understanding of different aspects.

3.2 DESCRIPTION OF PROCESS MINING

This mining technique is based on automated discovering and detecting methods and procedures from event logs. The process model is constructed based on observing events which means all the activities being executed, or the exchanged messages in a process. The most essential problem in process mining methods is one cannot handle all possible situations and behaviors. The current available process mining techniques are not suitable for different organizations and that consider an important issue because the aim of those organizations, is building a structure to detect logs. However, studying the different behaviors and making the model general helps the existing process mining technique avoiding that issue, in addition, it is done when implementing assumptions about completeness. It is observed, there is no existing technique that enables between what has been observed over-fitting and the more behaviors under-fitting (Van der Aalst et al., 2010). All the mentioned information about process mining emphasizes on what is called event log that is data recorded about processes in a specific domain in the industry. There are variety of "Process-Aware Information Systems (PAISs)" (Dumas et al., 2005) that can record excellent data on actual events in that domain. There are many examples of such system as Enterprise Resource Planning (ERP), Customer Relationship Management (CRM), Workflow Management (WFM), Supply Chain Management (SCM), and Product Data Management (PDM) systems (Dumas et al., 2005). Although, event logs are rich with information, most software vendors deploy this information to handle basic inquires that are assuming a well defined constant process. Therefore, process mining techniques aim to detect useful data out from system's files, and a control flow discovery which is one aspect of process mining that automatically constructs a process model that is describing the causal dependencies between activities (Van der Aalst et al., 2004b). The main idea of the control-flow discovery is based on constructing a suitable process model automatically by having a log file that contains a set of traces to describe the activities' behavior as presented in that log. For example, some algorithms like the "α-algorithm" (Van der Aalst et al., 2004b) structures a process model a "Petri Net model" by the identifications of classified patterns among the events of a log, such that one event from those events should follow another one. Despite of the advantages of using workflow models "WFM" in the design construction, but it needs a complete events log that contains structured data. However, in real life situations events that are recorded in those logs are not structured very well, hence many algorithms produce noisy diagrams which do not provide valid process models that contain deadlock, etc. (Agrawal et al., 1998; Van der Aalst et al., 2004b). Handling the different issues pf event logs information need more than one mining approaches, therefore, there are many different process mining methods differ from non global approaches that check the internal relations among activities in logs such as the α, $\alpha + +$, "Heuristics Miner" algorithms, while the non local approaches are modeling data files by implementing the whole log such as the "Genetic Mining, Region Mining, and Fuzzy Miner". Different algorithms have their own specialisms, e.g., α is proven to be able to mine models that follow the restrictions of Structured Workflow Nets (SWF-nets) (Weber et al., 2011) but cannot mine implicit dependencies or even handle noisy logs. However, the "Heuristics Miner" deploys different techniques to manage noise such as pasteurization and frequencies, while the "Genetic algorithm" mines complicated logs that even are noisy but are resource intensive. Nowadays, researches focus on developing new techniques that can manage complex real-world models or noisy logs. In this chapter, the famous process mining algorithms are described and the idea of workflow and how can be implemented in mathematical format is also presented. Next section is about Petri Nets and how they are implemented with process mining.

3.3 PETRI NETS

"A well known technique that is called Petri Nets (Petri, 1962), and it is defined as a graphical language for the representation of a process. A Petri Net is a bipartite graph, where two types of nodes can be used: transitions and places. The transitions nodes represent activities to be executed, and places nodes represent states whether an intermediate or final that the process can reach. While edges, are directed, and must connect a place and a transition, hence an edge is not allowed to connect two places or two transitions. Each place can contain a certain number of tokens where the distribution of the tokens on the network is called marking. In Figure 3.1, a small Petri Net is shown; circles represent places, squares represent transitions" (Devillers and Antti, 2015).

In figure 3.1, T means transition and P means Place. Knowing that Petri Nets are presented according to various perspectives to cover both the "clear semantic" and the "certain number of possible extensions", as an example color and time, ...

"A formal definition of Petri Net, as presented, for example, in Kononenko (1993), is the following:"

Definition 1 (Petri Net). A Petri Net is a tuple (P,T,F) where: P is a finite set of places; T is a finite set of transitions, such that $P \cap T = \emptyset$, and $F\ (P \times T) \cup (T \times P)$ is a set of directed arcs, called flow relation.

However, some basic workflow templates that can be modeled using Petri Net notation (Hakeem et al., 2012), and it is presented as Workflow (WF) Nets. WF Nets are the most famous graphical presentation used by some process mining algorithms because the WF Nets or the Petri Nets allow concurrency between events to be modeled explicitly, this is feature is very important because concurrency is critical part in business processes. The various types of Petri Nets help business developers to present process models in more details that allow those developers to apply improvements and avoid many problems that exist in the current model (Mani et al., 1997).

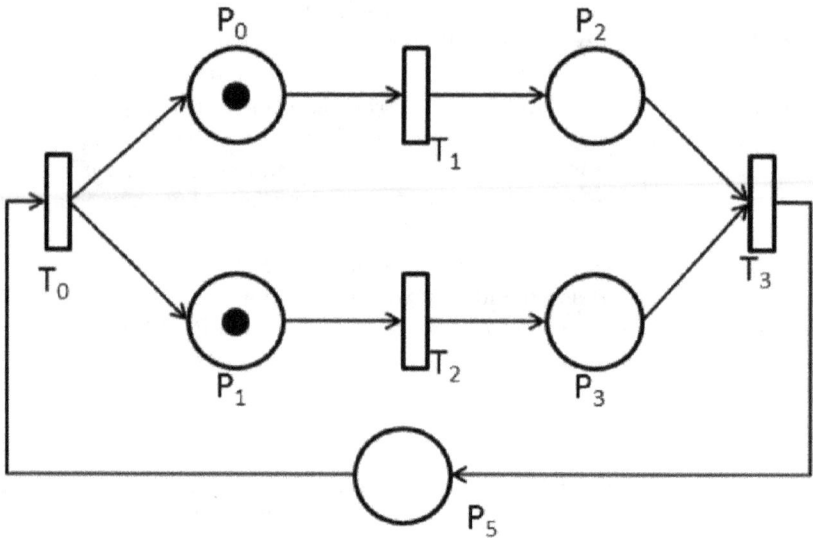

FIGURE 3.1
Petri Net Example

3.4   ALGORITHMS SURVEY

As it will be seen in this chapter, the α algorithm handles concurrency in logs of events. "It is proven to correctly mine processes where the underlying processes can be modeled by a structured workflow net (SWF-Net), a subclass of Petri Net. The algorithm is simple to apply. The α algorithm works as follows. First, it examines the log traces and creates the set of transitions in the workflow. Then the set of output transitions of the source place, and the set of the input transitions of the sink place. In the final two phases of α-algorithm creates sets of source places, and input places respectively used to define the places of the discovered workflow net. In phase 4, the α-algorithm discovers which transitions are causally related. Thus, for ant two events in the workflow, each transition in the first set causally relates to all transitions in the second set, and no transitions within the first set or the second set follow each other in some firing sequence. These constraints to the elements in the two sets allow the correct mining of AND-split/join and OR-split/join constructs. Note that the OR-split/join requires the fusion of places. In step 5, the α algorithm refines the first set of events by taking only the largest elements with respect to set inclusion. In fact, Step 5 establishes the exact amount of places the discovered net has excluding the source place and the sink place. The places are created in Step 6 and connected to their respective input/output transitions in Step 7. The discovered workflow net is returned in Step 8. Finally, it is defined what it means for a workflow-net to be rediscovered. Although α is considered unsuitable for real-world processes, is often used to obtain a first insight into a process. α requires a complete event log as I have published in article of Zayoud et al. (2017b), the meaning of the term completeness of a log file is if there are pair of events in the model, hence according to the rules of that model those events are allowed to directly follow each other and this means to have proof of that behavior exists in the log file. However, the α depends on this behavior to extract process model and in case of noisy log files the α algorithm is failed to handle such a problem which is the noise in a log file (Alves de Medeiros et al., 2004a).

The internal structure and the representation of individuals that constitute a population are considered as the constitution of the genetic material of that population. The crossover and mutation (and/or) operators that are related to genetic can modify the behavior of genetic nature of some parts as an example, the operator of crossover which merges between individuals that consider as parents of a population to come up with new offspring or even individuals of the coming generation. While, the other operator called mutation can modify "parts of individuals in the population based on random procedure" (Zayoud et al., 2019a). For the two operators, there is a selection criterion for the suitable weather for the "crossover" and/or "mutation". It is important to know that, in any population, the number of the best individuals that is named as the elite of the population guarantees that saving of the good genetic material to be not lost instead it is directed to the next generation as a copy. However, the creation of generations is continues using a continues search process in new populations until achieving a required criterion, but sometimes, the search process ends even if that search did not meet the required criterion because of setting a biggest number of generations with the latest generated populations while implementing a search technique that is offered by the genetic algorithm. Then, ending without finding individual with maximal fitness (Alves de Medeiros, 2006).

Another algorithm belongs to process mining algorithms is called the Heuristics Miner algorithm that is a practical applicable algorithm to handle the noise in a log file by using only the basic behavior without the need of detailed information or exceptions recorded in an event log. This algorithm can control any process model with its flow perspective and by taking into consideration activities' order in each situation, by depending on this fact, to neglect their order through cases, hence the causal dependencies should be analyzed based on log events as a method of finding a process model. Such as, if two events are co-related by follow relationship so when there is order one event should start before the other one, this means there is a high possibility to find a dependency relation for those events. This causal dependency is the building block in the Heuristics Miner and the basic of constructing what is called a dependency graph. Another important factor in the Heuristic Miner is a frequency matrix and its aim is presenting a dependency relation that may appears between events in a log file such as an event is named "A" and another one named "B".

A threshold point is achieved in a heuristic method if a correct dependency relation is detected between events in an events log. The correct dependency relation means to have fellowship between some events in the log file such as one event should be followed by another event in the same log file for many times, and its value of this fellowship relation is between -1 and 1. This range of the values means, the amount of dependency between event "A" and "B" after monitoring different traces of those related events, hence the value is 1 if event "A" is followed by event "B". The relation of causality between events in a noisy log can be found using this simple heuristic enormously, and the model that is usually implemented to construct an event log is known for all the traces weather the normal traces or the ones with noise. However, in the real life situations it is difficult to differentiate between the trace with noise or the trace that is a low frequent pattern, hence three threshold parameters should be exist in the "Heuristics Mine" as shown in Zayoud et al. (2017b):

- the threshold of dependency,
- the threshold of positive observations, and
- the threshold of relative to best.

The mentioned thresholds provide indications if there is also an accepted relation of dependencies among events under the following rules:

- the value of a dependency that is measured should be higher than the "Dependency Threshold",
- the value of frequency level should be greater than the "Positive Observations Threshold" measurements, and
- the calculations of subtraction between the best measurement of dependency and the measured value of dependency should be less than value of "Relative to Best Threshold".

"The heuristic miner as shown can't handle short loops, and the type of the dependency relations (AND/XOR-split/join) isn't represented in the dependency graph, in addition there are problems with non-observable activities, and it can't handle long distance dependencies" (Weijters et al., 2006b).

Fuzzy mining is a process discovery technique that mines an event log for a family of process models using a map metaphor. As many maps exist that show a specific city as example at different levels of abstraction, also different maps exist for a process model mined from an event log. In this map metaphor is an object of interest in that city (like specific museum) corresponds to a node in the process model, where streets (like street 1 for example) correspond to edges in the model. A single map is a fuzzy instance whereas a family of maps like all the city maps is a fuzzy model. Like high-level maps only show major objects of interest and major streets, high-level fuzzy instances show only major elements nodes and edges. For this purpose, the Fuzzy Miner computes from the log a significant weight for every element and an additional correlation weight for every edge. The higher these weights are, the more major the element is considered to be. Furthermore, the Fuzzy Miner uses a number of thresholds. However, only elements that meet these thresholds are shown. These thresholds correspond to the required level of abstraction: the higher these thresholds are, the higher the level of abstraction is. To reach completeness, it should be mentioned that a fuzzy instance may contain clusters of minor nodes: If some objects of interest on that city map are too minor to be shown by themselves on some map, they may be shown as a single and major object provided that they are close enough. For this reason, the Fuzzy Miner first attempts to cluster minor nodes into major cluster nodes, and only if that does not work it will remove the minor node from the map. The fuzzy model view, allows the user to change the thresholds by changing the sliders whereas the fuzzy instance view does not which I have published in Zayoud et al. (2017b). The significance's consideration of any point is depending on some features that describe that point or node, for example, if "image 1" is a point significance's value equals to 0.253, while for the "edge" the significance's value is seen by checking the wideness level of that edge and it this width increased, hence its significance level is also increased. The wideness is not the only feature of that edge, also its color brightness provides an idea about the level of "correlation" if the color is brighter hence it will be less correlation to its input point or node in addition to its results or output point (dos Santos Garcia et al., 2019).

• Process Mining Algorithms

This part is explaining the main algorithms of "process mining" are presented in a type of survey about their important applications.

"Nowadays, information systems contain event logs that are distributed among multiple devices. Hence Map-Reduce is considered as a salable procedure for efficient computations of distributed data (Evermann and Assadipour, 2014), and there is a correlation among the "α process mining algorithm" and a Map-Reduce in terms of its scalability advantage which can improve the performance of the "α" implementation especially when they are implemented as experimental purposes. However, Log of events which have valuable information about the system with its different events especially what is called a "process instance", this instance is defined as an activity "type" in a case the task is with a "timestamp", and implementing "α process mining algorithm" is for rediscovering a possible "Workflow Net" of a log assuming a complete and free of noise. "Knowing that, Map-Reduce is an approach that handles large scale of data that can be considered as big data because of its data distribution feature, in addition to its ability to follow basic procedures like the "map function" which is a function which takes parameters as pairs of sequences and returns also pairs as output results, also another function which is called a "shuffle function" that gathers items that have different keys to present the results as sequence of "tuples". Then a "reduce function" uses these tuples an inputs to provide items that are presented in sequence of "key pairs". As an advantage of applying a "Map-Reduce" based algorithm, is the combination of balanced sequences of "map and reduce" functions all merged with suitable "keys and values" in reference to their types of data. To compute the correlations and organize them into a log using the α algorithm, it is required two groups of "mappers" and "reducers", that are related by fellowship as seen in Zayoud et al. (2019a).

Proceeding with this survey, "the Fuzzy Miner" (Günther and Van Der Aalst, 2007a) which is defined as a "dynamic" approach of what is called thematic cartography methods, which mainly depends on the generalization concept as a basic procedure, it means the abstract of certain details to be changed frequently. Knowing the fact about "fuzzy miner" which the identifications of two matrices: significance as the first one and the second one is the correlation. Significance metric computes the frequent events and their orders, i.e., the more frequent a priority relation is seen, hence a higher significant it is. In reference to "correlation" matrix which is about measuring the strong relations between events in a log and tested by the amount of common data, or by checking the names of events if they are similar, both metrics are of a model for a process depending on the level of significance and correlations between events of a log, hence if significant behavior is clearly seen this also means the correlation behavior is strong as well. On other hand, when significance is weak to be observed, or events are aggregated into clusters, or not clustered hence the weak behavior of correlation will be resulted. All the mentioned facts about the affects of the two matrices helps to build what is called "dynamic views" of events in a log which are powerful to learn more about the process in order to achieve an accurate model of that process. The mentioned approach is deployed in an application that is related to "internal transaction fraud mitigation in well-known company, such that the process diagnostics only concerns on a global view of the business process, in order to help the analysts and domain experts to reveal weaknesses and problems in the business process and discovers any fraud cases may happen and exposes opportunities to commit fraud in the followed process as seen in the paper" (Zayoud et al., 2019a). The next sections in this chapter is a detailed explanation of process mining and how to select the best one then an explanation of big data concepts and how can be applied in the industry espcially in healthcare systems.

3.5   PROCESS MINING ALGORITHMS SELECTIONS

There are various proposal for algorithms to mine processes, but it is not possible to have an approved procedure to judge and check these mining algorithms, and as a result, choosing a suitable method for a given enterprise it is a difficult task. The evaluation of those process mining algorithms is implemented on the outputs of those algorithms that are presented as process models that should be checked with the primary models of the mined organization for conformance feature in terms that discovers more efficient and streamlined business process models that are semantically equivalent to the original process models to keep the existed resources. "However, process mining algorithms have different features and some of them perform better on models with unkown tasks, while other algorithms do not, hence being able to select the most appropriate process mining algorithm means to choose the best one that produces mined models and those process models are similar to the original models and structurally equal to or better than the original models and recent research in addition to software prototypes have attempted to provide such an evaluation framework" (Zhou et al., 2012). "The evaluation of available process mining algorithms for the business models provided by a given enterprise is expensive and time consuming especially regarding the business process models that are evolving rapidly. Therefore, these process model mining algorithms should be re-evaluated regularly against these changing models for conformance checking, reengineering, or discovery of more streamlined, improved models. The basic goal of this chapter is to have the knowledge to evaluate, compare and rank these process mining algorithms efficiently. Also to investigate how to choose an effective process mining algorithm for an enterprise without evaluating different process mining algorithms extensively.

Moreover, to know how to avoid re-evaluating all the algorithms whenever the business processes of the enterprise change" (Wang et al., 2012a). The process mining algorithms being evaluated in this chapter are: α algorithm (Weijters et al., 2006a), genetic algorithm (Wu and Kumar, 2009), heuristics miner (Duda et al., 2001), and region miner (D'Alessandro et al., 1999). Without loss of generality, these four algorithms represent the four most popular classes of business process mining algorithms.

• Process Mining Using Heuristics Miner algorithm

The "Heuristics Miner" algorithm aims to detect the workflow from the perspective of events in the system based on their model. To control the flow of any process model, events' order in a particular case should be considered without their order among cases because in any log file, some fields are necessary such as the id, and the timestamp of an event. Those fields are important because they are playing a critical rule in specifying some relations between events in a log file especially those relations that can be recognized by knowing the order of events. Moreover, the detection of a process model requires an analysis of log events in reference to "causal dependencies" between those events, if one event is followed by the other event, hence there is a huge possibility of dependent relation between the two events. To analyze these relations, some notations are produced and defined to apply these rules (Weijters et al., 2006a).

• Algorithm Description

The first phase of "Heuristics Miner", is building a "dependency graph" based on a "frequency metric" in addition to the indication of a "dependency relation" between events like "A" and "B" (notation $A \Rightarrow WB$). The calculated values of an "events log are used in a heuristic search for the correct dependency relations" (Weijters and Ribeiro, 2011). If "W be an event log over T", "and $a, b \in T$. Then $|a > Wb|$ is the number of times $a > Wb$ occurs in W, and" (Weijters and Ribeiro, 2011)

$$A \Rightarrow_W b = \left(\frac{|a >_W b| - |b > Wa|}{|a >_W b| + |b >_W a| + 1} \right)$$

"First, the level of $a \Rightarrow Wb$ is ranged between -1 and 1. Some examples demonstrate the rationale meaning behind this definition. If this definition is used in the situation that, in 5 traces, activity "A" is considered as a "direct follow" by "B" but not the opposite, the value of $A \Rightarrow WB = 5/6 = 0.833$" (van Dongen et al., 2007a) indicating that there is no guarantee completely the "dependency relation (only 5 observations possibly caused by noise). However, if there are 50 traces in which A is directly followed by B but the other way around never occurs, the value of $A \Rightarrow WB = 50/51 = 0.980$ indicates that there is pretty guarantee of the dependency relation. If there are 50 traces in which activity A is directly followed by B and noise caused B to follow A once, the value of $A \Rightarrow WB$ is $49/52 = 0.94$ as presented in the paper" (Weijters and Ribeiro, 2011) indicating that a dependency relation is guaranteed.

"A high $A \Rightarrow WB$ value strongly suggests that there is a dependency relation between activity A and B. But what is a high value, what is a good threshold to take the
decision that B truly depends on A. The threshold appears sensitive for the amount
of noise, the degree of concurrency in the underlying process, and the frequency of
the involved activities. However, for many dependency relations it seems unnecessary to use always a threshold value. After all, each non-initial activity must have
at least one other activity that is its cause, and each non-final activity must have at
least one dependent activity. Using this information in the so called all activities-
connected heuristic, it can take the best candidate (with the highest $A \Rightarrow WB$ score)
(van Dongen et al., 2007a). This simple heuristic helps us enormously in finding reliable causality relations even if the event log contains noise" (Zayoud et al., 2019a).
"As an example, having a noisy log of events, hence the heuristic approach is possible to be applied with "Petri Net" as well. Also, an example of 30 traces of events
where half of them are possible and other half is not right: AD, $AECBD$, $ABCED$
from the $\log W = [AECBD, AED9, ABCD9, ACBD, ABCED, AD]$. The initial part
is to get the $\Rightarrow$-values for all valid combinations of activity. The result is expressed
in the matrix" as shown in Figure 3.2.

$\Rightarrow_W$	A	B	C	D	E
A	0.0	0.909	0.900	0.500	0.909
B	0.0	0.0	0.0	0.909	0.0
C	0.0	0.0	0.0	0.909	0.0
D	$= 0.500$	-0.909	-0.900	0.0	-0.909
E	0.0	0.0	0.0	0.909	0.0

FIGURE 3.2
Activity Cominagion Matrix
Source: (Xue and Zhu, 2009).

The starting activity A in the matrix is determined as non-positive value in the
A-column by deploying the "all-activities-connected heuristic", while the dependent
activity A is determined by searching for the largest number in row A of that "matrix". Looking at the hight amount of B and E which are (0.909), hence "B" will be
chosen arbitrarily. In reference to causality in the same matrix, we need to search
for B as the maximum amount in B column, which helps to detect the fact of A as
the cause of B. Moreover, D and B are related with dependency relation where the
D is dependent on B because D activity has the biggest one in the B row. Following
similar approach on other activities such as B, C, and E; determines only the causal
relations that are described in what is called a "dependency graph". In fact, there
are values in the activity cells present the activity's frequency, while the number of
how many arcs describes the constant causality relation of events and their frequent
behaviors. If noise in a log is ignored, then "causal relation" can be mined efficiently. In the presented model it can be seen from the process which is implemented

previousely to establish the log of events that are traced with existance of noise (i.e., *AECBD*, *AD*, *ABCED*), but in the real applications in life are never known if for instance the "trace" AD wheather it is a "low frequent pattern" or has a "noise". To handle this, the "HeuristicsMiner" has three available parameters named as threshold which are for "Dependency", "Positive Observations", and "Relative to Best" thresholds. The measurements of those thresholds are indication of a dependency among events according to: (i) its level if higher than the "Dependency threshold", (ii) having more frequent than the value of the "Positive Observations threshold", and (iii) if the difference among the "best" is less than the value of "Relative".

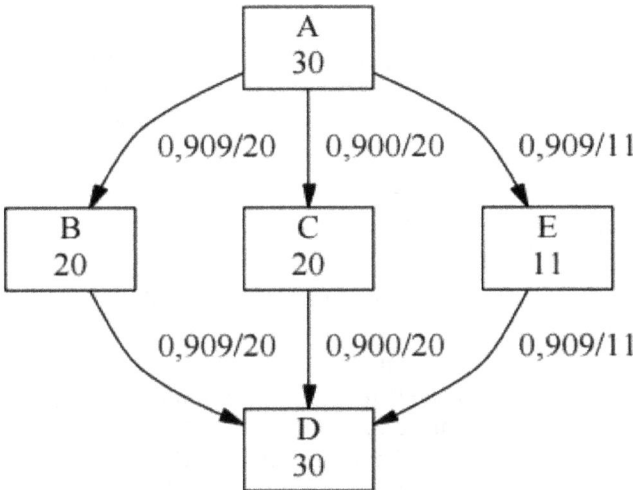

FIGURE 3.3
Implementing heuristic algorithm on noisy logfile to get a dependency graph using Petri Net of simple example.

In the figure above, a dependency graph is resulted from the settings of the threshold parameters. *AD* is presented in the graph becasue of its low frequency with a "Dependency threshold" $= 0.45$, "Positive observations threshold" $= 1$, and "Relative to best threshold" $= 0.4$. However, the real situations that usually are not ideal in terms of complete events logs that may have hunderds of traces, noisy, or even traces with low frequncy level, can be analyzed and studied using the three mentioned treshold parameters to identify any effective behaviors of processes. Eventhough, those parameters are powerful but the basic algorithm as presented above is not considered the suitable algorithm to deal with "short loops", or any other logic relations that are not even described in the dependency graph, the non-observable activities, or long terms dependencies, but in reference to loops that are not long in a process, it will be executed short loops in a process. However, as an example of long terms of loops (e.g., ...*ABCABCABC*...) which have no issues with the Heuristics Miner presented so far (the values of $A \Rightarrow WB$, $B \Rightarrow WC$, and $C \Rightarrow WA$ are useful to indicate dependency relations). However, for length-one loops (i.e., traces like *ACB*, *ACCB*,

ACCCB, ... are possible) and loops of length two (i.e. traces like *ACDB*, *ACDCDB*, *ACDCDCDB*, ... are possible) the value of $C \Rightarrow WC$ and $C \Rightarrow WD$ is always very low. However, it appears very simple to define the dependency measure for loops of length one and length two. Let W be an event log over T, and $a, b \in T$. Then $|a > Wa|$ is the number of times $a > Wa$ occurs in W, and $|a >> Wa|$ is the number of times $a >> Wb$ occurs in W

$$a \Rightarrow_W a = \left(\frac{|a >_W a|}{|a >_W a| + 1} \right)$$

$$a \Rightarrow_{2W} b = \left(\frac{|a >>_W b| + |b >>_W a|}{|a >>_W b| + |b >>_W a| + 1} \right)$$

During the construction of the dependency graph loops of length one is treated in the same way as other activities (they need an external cause and a dependent activity). Loops of length-two need a special treatment while applying the all activities-connected heuristic. They form a pair and as a pair they need only one cause and one depending activity (Weijters et al., 2006a).

• Process Mining with the Genetic Algorithm

"This section presents a process representation, a method to measure and the genetic operators used in a genetic algorithm to mine process models. The focus is on the use of the genetic algorithm for mining noisy event logs. Genetic algorithms are considered adaptive search methods that is used to mimic the process of evolution (MacQueen et al., 1967a; Yeung et al., 2007a). This algorithm starts with an initial population of individuals which is the process models. Populations evolve by selecting the fittest individuals and generating new individuals using genetic operations such as crossover by combining parts of two of more individuals and mutation random modification of an individual. The initial experiences show that a representation of individuals in terms of a Petri Net is not a very convenient because the Petri Net contains places that are not visible in the log. For example, in Figure 3.4 meaningful names to places cannot be assigned. Second, the classical Petri Net is not very convenient notation for generating an initial population because it is difficult to apply simple heuristics. Third, the definition of the genetic operators i.e., crossover and mutation are cumbersome. Finally, the expressive power of Petri Nets is in some cases too limited in combinations of AND/OR, splits/joins" (Ramakrishnan, 2009b). Hence, what is called a "casual matrix" has to be deployed.

Any model of a process should present existed activities for that business process. The presentation of those activities or events is done by having a route, this route is used to recognize any activity and if it is a strong cause for other activities. The route is considered as sequential as in Figure 3.4 sequence if any activity is single cause of any other activity. But the route is considered as parallel as in Figure 3.4 – parallelism if an activity allows the implementation of many instantaneous activities. While the route is considered as choice if an activity allows an implementation of various events with one of them must be done only as in Figure 3.4 – choice. These routing considerations i.e. sequence, parallelism, and choice may merge to propose

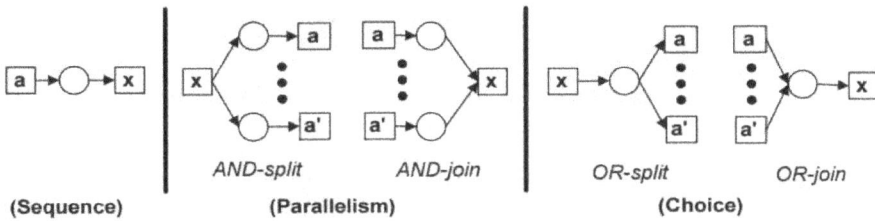

FIGURE 3.4
Petri Net for modeling business processes for the three basic routing constructs.

a complex model such as loop which is known with the implementation of a "sequence" and a "choice where the OR-join precedes the OR-split". Monitoring those routing constructs, hence the model of a process has to present initially the existed events in the process, second "cause/enable" relation between events, third checking what is detected weather "causal" relation among events in case of sequential combinations, or even "parallel" or "choice" routing. To understand the design of a process model and how was established, we need to know about matrix that has various rows and columns with Boolean terms. This link between process model and matrix is to present the existence of "causal relations" ($\rightarrow$) between activities in that process, hence it is named as "Causal Matrix" and the length of it equals to count of activities of a process and this length is expressed as "$n \times n$". Those activities are helpful to express the routing constructs. The contents of this matrix is only Boolean values to present the outputs of the logical operations "AND/OR-split/join" that are actually the types of those relations in that matrix with only ($\wedge$) and or ($\vee$) as Boolean operators.

The main definition of individuals is achieved by explaining how it is encoded in the Genetic algorithm. It begins with knowing the input as events and output as "Boolean expressions" which are the direct derivative of a complete causal matrix that is not stored precisely and is used only in the phase of initializing a process. The output sets of 1s and 0s entries to the matrix and map to sets of subsets. Subsets with events that have an OR-relation and other subsets under AND-relation. For instance, the Boolean expression $(e \vee f) \wedge g$ equals the set representation $\{\{e, f\}, \{g\}\}$. The conceptual encoding shown in Table 3.1, that is, mapped to the implementation one. Knowing that Table 3.2 assumes a "normal form", i.e., "a conjunction of disjunctions can minmize the space with having some kind of limitations of the expressiveness, cf. In a log free from noise, the genetic search should achieve an optimal process that complies with the information in an event log, and that is make the Genetic approach attractive if the event log contains noise" (Ramakrishnan, 2009b).

• Process Mining with Region Algorithm

"The Theory of regions (Chapelle et al., 1999; Friedman, 1977a) establishes a connection between transitions systems and Petri Nets through net synthesis. The idea behind the Theory of Regions is that a state-based model which is a model

TABLE 3.1

Causal Matrix of an Individual in the Inherent Considerations

	Input								
T	a	a	a	d	d	$e \wedge f$	$b \vee c \vee g$	
$\rightarrow$	a	b	c	d	e	f	g	h	Output
a	1	0	0	1	1	0	0	0	$a \vee d \vee e$
b	1	1	1	0	0	0	0	0	$a \vee b \vee c$
c	0	1	0	1	0	0	1	0	$b \vee d \vee g$
d	0	0	0	0	1	1	0	0	$e \wedge f$
e	0	0	0	0	0	0	1	0	g
f	0	0	0	0	0	0	1	0	g
g	0	1	0	0	0	0	0	0	b
h	0	0	0	0	0	0	0	0	T

describing which states a process can be in and which transitions are possible between these states, can be transformed into a Petri Net, in a compact representation of the state space, explicitly showing causality, concurrency, and conflicts between transitions.

The Theory of Regions shares common goals with the research area of process mining. However, there are some differences:

- First, the starting point for net synthesis is called transition system which is a description of a process explicitly showing all possible states, whereas event logs do not carry state information.
- Second, the theory of regions assumes the transition system to show all possible transitions between states, while in process mining, the assumption usually is that the logs are not exhaustive, and do not contain all possible sequences of events.

Regions and the related theory has been developed and successfully applied to what called net synthesis (see, among others, Gray and Neuhoff, 1998; Lloyd, 1982; Quinlan, 1986) and to the characterization of concurrency models (see, among others, Cristianini et al., 1999; Huang et al., 2004; Smola et al., 1999). However, the approach of process mining based on regions has not received yet great attention. In fact, the use of regions gives in general a saturated net (i.e., with many more places) and the complexity is in general quite high in comparison with other method. Moreover, the novelty of this method relies on the incremental calculus of regions. Though regions of a transition systems can be combined algebraically under precise conditions (Friedman, 1977a; Wu et al., 2008), the attempt to find regions of a compound

TABLE 3.2

A More Encoding of the Sample

"Activity"	"Input"	"Output"
A	$\{\}$	$\{\{B,C,D\}\}$
B	$\{\{A\}\}$	$\{\{H\}\}$
C	$\{\{A\}\}$	$\{\{H\}\}$
D	$\{\{A\}\}$	$\{\{E\},\{F\}\}$
E	$\{\{D\}\}$	$\{\{G\}\}$
F	$\{\{D\}\}$	$\{\{G\}\}$
G	$\{\{E\},\{F\}\}$	$\{\{H\}\}$
H	$\{\{B,C,G\}\}$	$\{\}$

transition system from the regions of the components is new, and this gives better performances. Moreover, that regions have been used in many different settings such as in the synthesis and verification of asynchronous circuits (MacQueen et al., 1967a) or in the verification of security properties (Hunt et al., 1966). As explained before, in process discovery, there is no model to start with, and the Theory of Regions is still highly relevant. Moreover, if there is a process log file, and a Petri Net that describes exactly what seen in that log is required to be obtained, hence the theory of Regions would apply directly. However, there are still some issues in using the Theory of Region related to the process log when process instances are sequences of events and do not carry any state information, so there is no relation between different process instances, and will be never known if the process log is large enough to exhibit all possible behavior of the underlying process" (Chapelle et al., 1999; Friedman, 1977a).

Region-based Process Discovery

The completeness concept is start step and assumption of Regions theory especially for a process log that will be used for the study. The output of this synthesis study appears as Petri Net that describes the flow of events as given in the log file, it should be assumed that the log shows all possible behavior, which means it is globally complete. Moreover, a transition system is needed when the theory of Regions is implemented because the log of events that represents the process should be converted to transitions system, hence what is called states must be recognized. One challenge of identifying states that in some cases some of those states may not be exist in log file. Facing this challenge needs a powerful method such as a Naive approach which works by considering only one known state and call it the initial state, also considering a unique activity in a log as the 1st one in that "Sequence" in tha file log.

However, it is not trivial to depend on the assumption of starting state that is unified for all traces as an "Intuitive one", and assuming also the start step all traces should be with a unified transition because in practical situations, traces simply can begin with various alternatives especially when customers are vary as either recurrent or new. It is guaranteed by providing an extra activity at the beginning of every trace to force having a restrictive assumption. The begining phase in a process of region mining is making every trace in a process log as a transition system. After previous assumptions it can be considered that the Theory of Regions has a great impact in discovering the process, and its output result in term of Petri Net regenerates the aimed log of events. Even though, the theory of Regions helps to achieve the required goal but it still has some disadvantages such as the knowledge of a log file in term of its completeness level. Moreover, the resulting Petri Net should provide better flexibility level to more behaviors than the current one. As a solution of this flexibility challenge is having a bigger log file which can give better results and more knowledge level but the other challenge whether a log file is not noisy and complete may not be answered. Another example about limitation of this approach is the transition system that should be established even before calculating regions, this issue is considered as a limitation because it is not feasible in term of the huge memory, i.e., the space allocation that is required in case of handling big and complex logs. To summarize the above ideas, all algorithms that are related to process mining have to take into considerations what is the most important fact the computation time or memory space restrictions. In Fayyad et al. (1996), as an example, the method of mining a process by the "Genetic algorith", this algorithm is used here because its feature of linear growth in the log size, the linear scale make this algorithm is good in term of the needed memory to handle a log file that has double number of cases but the it will be implemented with more time in suh situation, but requires the same amount of memory. However, when a full transition system is stored in memory, the calculation of only minimal regions is simpler than in our iterative approach, i.e., using a breadth first search, all minimal regions could be found, without considering larger regions. Our iterative approach requires all regions to be calculated, after which the minimal regions need to be found in the set of all regions. Therefore, the computation time is larger with our iterative approach. Finally, the "Theory of Regions" is used in identifying processes, by presenting loo file as an output of "Petri Net" term as:

1. Each trace in the log is a firing sequence in the Petri Net and
2. Each firing sequence in the Petri Net is a trace in the log.

The iterative approach that is applied by this algorithm may minimize space complexity to be commonly accepted as the bottle neck of various mining approaches for process. The hardest issue of the theory of "Regions" in the context of process discovery is associated with the advantage in the theory of Regions domain. The output Petri Net describes the workflow of events in any log accurately, therefore, the Theory of Regions is aimed to produce compressed representation of an initial transition system as a Petri Net output, which is considered different from the idea of using process mining, that mainly means generating a precise model for the process in that log (Chapelle et al., 1999; Friedman, 1977a).

• Process Mining with the α-Algorithm

The α-algorithm is a mining procedure to re establish "causality" among sequences of activities. It was first put forward by "Van der Aalst, Weijters, and Maruster" (Van der Aalst et al., 2004a). It is one of the basic process mining algorithms therefore, it has several extensions or modifications. "The α algorithm is designed to handle concurrency in processes and proven to correctly mine processes where the underlying process can be modeled by a Structured Workflow Net (SWF-Net), a subclass of Petri Net. This algorithm is simple to apply, but it is not suitable for processes in real life because it is deployed to achieve a first view of a process only. Moreover, the "α algorithm" needs log of events that is complete, where completeness means the direct follow for each pair of events that exist in a process model, and this is found as a trace in the log that "exhibits this behavior. It uses such pairs of activities identified from the event log, to attempt to exactly replicate the underlying process from the traces in the event log (Zayoud et al., 2019a). Therefore, it is unable to handle noise, informally, noise-free logs are defined as those recorded without error, by a process which is followed without error. It is noted that process, system and data problems may cause α to produce a difficult to understand, or "spaghetti" Lerman (2000) model. Heuristics Miner is explicitly designed to handle noise in event logs, and for this reason has been the algorithm of choice in many practical applications (Agrawal et al., 1995), where processes have been found to be complex and poorly recorded. Examples are local government Lewis et al. (2004), healthcare (Wang et al., 2012b), finance" (Dempster et al., 1977) and telecoms (Friedman, 1977a). Whereas α decides any result based on the existence or the variation in counts of some pairs of activities in a log file, the "α" uses this procedure to recognize which arcs that should be established in the model. However, the α algorithm is presented in details in the next section (Van der Aalst et al., 2004a).

• α-Algorithm

The α-algorithm is the startup mining process method to achieve workflows nets as "Petri Nets" as outputs from log of files.

The discovering process of a workflow contains the follwoing phases

- "pre-processing": analyzing relations in transitions
- "processing": implementing α approach
- "post-processing"

Process mining is the main topic of my study through the doctorate degree and it is explained in through this book, and Figure 3.5 presents the different aspects of process mining.

• Workflows Examples

As an example of workflows, Figure 3.6 shows a workflow that does not contain any implicit places where implicit places where having or missing them will not have an impact on the workflow behavior.

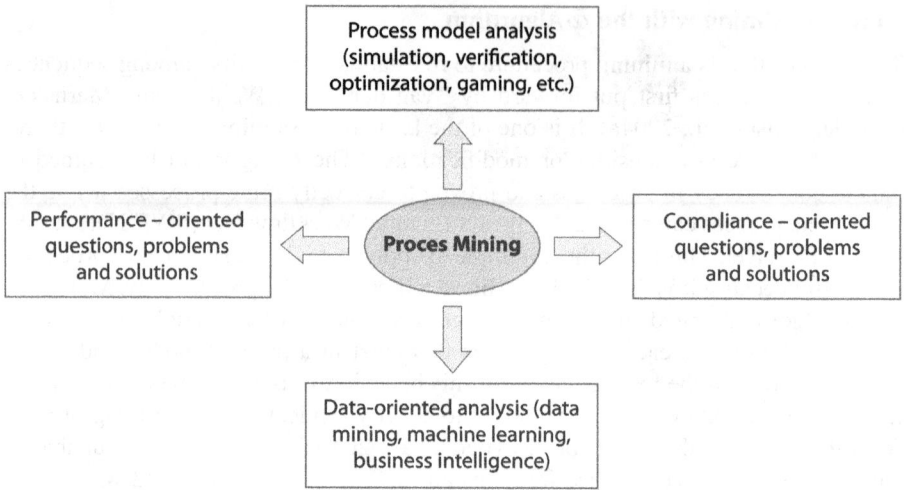

FIGURE 3.5
The process mining positioning meanings
Source: (Van der Aalst et al., 2004a).

Figure 3.7 shows that but if a there is a connection between "a and d" hence a new place is established, and it will be implicit, and will be invisible in any log due to unseen effects coming from them (Van der Aalst et al., 2004b; van Der Aalst, 2016).

● α **Algorithm Explanation**

Events logs are building block of the "α algorithm" to supply a workflow-net, where it is mandatory to find the ordering between transitions in a workflow, and by the recognized relations will be possible to get the "places" and any contact between them and the transitions of that workflor.

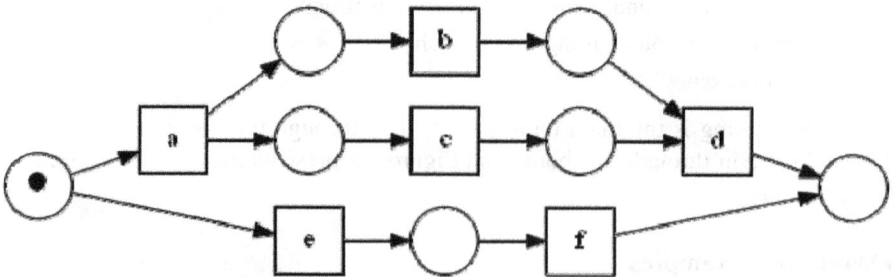

FIGURE 3.6
Workflow example without an implicit place
Source: Van der Aalst et al. (2004b); van Der Aalst (2016).

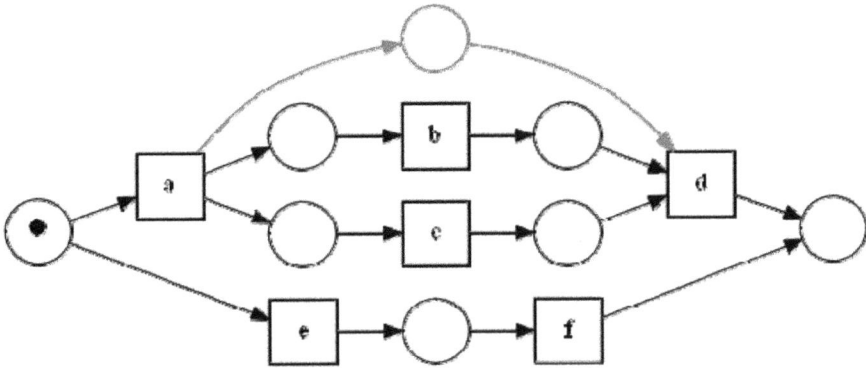

FIGURE 3.7
Workflow with implicit places
Source: Van der Aalst et al. (2004b); van Der Aalst (2016).

For example, the following is
"Transitions" and their relations (Van der Aalst et al., 2004b):
- *"direct succession"* x > y
  - "$x > y \Longleftrightarrow$ in log sub-traces $...xy...$"
- *"causality"* $x \rightarrow y$
  - "$x \rightarrow y \Longleftrightarrow x > y$" $\wedge y \not> x$
  - i.e. if there are traces $...xy...$ and no traces $...yx...$
  - this relation may mean that we will need to put a place between x and y
- *parallel* $x\|y$
  - $x\|y \Longleftrightarrow x >= y \wedge y >= x$
  - i.e. can see both $...xy...$ and $...yx...$
  - cannot put a place for such x and y – if we placed, we'd impose some order on them
  - this is symmetric relation "$(a\|b \rightarrow b\|a)$"
- *unrelated* x#y
  - $x\#y \Longleftrightarrow x \not> y \wedge y \not> x$
  - i.e. there are no traces $...xy...$ nor $...yx...$
  - this is also symmetric relation $x\#y \rightarrow y\#x$

The set of all relations for a log L
- is called the *footprint* of L

For example, in the Figure 3.8. Below the relations between transitions are defined
as
- the log is $L1 = [abcd, acbd, ef]$
- $a > b, a > c, b > c, b > d, c > b, c > d, e > f$
- $a \rightarrow b, a \rightarrow c, b \rightarrow d, c \rightarrow d, e \rightarrow f$
- $b\|c$

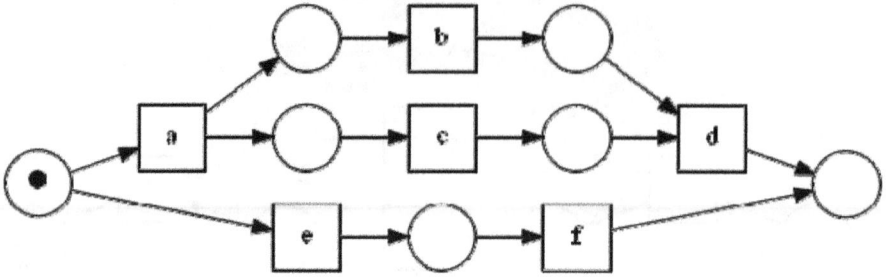

FIGURE 3.8
Workflow example

Petri Net model is provided in Van der Aalst et al. (2004b) and van Der Aalst (2016) to describe any event log in which is formalized as what is named α algorithm.

"The relations between any two activities, for example, a and b in a log are as following:

1. the direct succession $a > b$, when, in the log, sometime "a" compares before "b";
2. the causal dependency (or follow) $a \rightarrow b$, when $a > b$ and $b \not> a$;
3. the parallelism $a\|b$, when $a > b$ and $b > a$;
4. Uncorrelation #, when $a \not> b$ and $b \not> a$.

Given sets containing all the relations observed into the log, it is possible to combine them generating a Petri Net, following the rules presented in Figure 3.9" (Burattin et al., 2013).

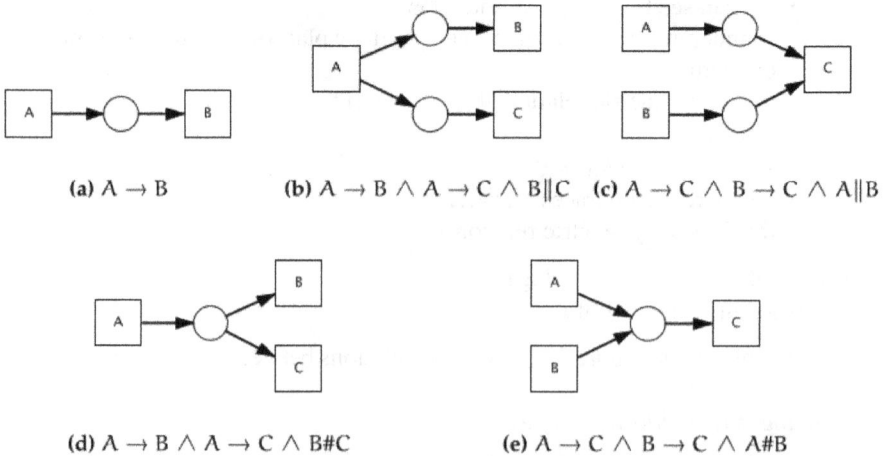

(a) $A \rightarrow B$ (b) $A \rightarrow B \wedge A \rightarrow C \wedge B\|C$ (c) $A \rightarrow C \wedge B \rightarrow C \wedge A\|B$

(d) $A \rightarrow B \wedge A \rightarrow C \wedge B\#C$ (e) $A \rightarrow C \wedge B \rightarrow C \wedge A\#B$

FIGURE 3.9
Petri Nets rules
Source: Van der Aalst et al. (2004a).

• Process Mining Using α Algorithm

The process mining is illustrated as a sample event log is taken into consideration as shown in Figure 3.9 where the log serves as the step one.

There are basically three different perspective for process mining.

These are (Banerjee and Gupta, 2015):

a) The process perspective ("How?").

b) The organizational perspective ("Who?")

c) The case perspective ("What?")

The main goal of process perspective is to focus on the control flow of the entire process by showing the steps included in the process in a sequential manner. The entire ordering of the events is shown with the help of Petri Nets (Reisig and Rozenberg, 1998) or Event Driven Process Chains (Keller and Teufel, 1998). Organizational perspective targets to the originator field from the log table. It shows that which person is involved in that process and how they are related. The result of this perspective is to differentiate people in terms of the organizational structure i.e. roles and responsibilities (Moreno, 1934; Nemati and Barko, 2004; Scott, 2012; Wasserman et al., 1994).

α algorithm is one of the basic 8-step algorithm that is deployed to mine processes. Its goal is to re establish "causality" of activities in a data file. Petri Nets (Reisig and Rozenberg, 1998) (also known as "P/T" nets) with distinct characteristics is produced as the output of this algorithm from event logs. Any log generating business management tool like an ERP system, etc. provides the event logs required. The net consists of events and transitions from events, i.e., the task being performed over the events. The entire Petri Net depicts the entire process as a process model that is recorded in that log. The input for the algorithm is file log of a workflow "$W \subseteq T^*$" that is expressed as a net to be the output of this algorithm, where T is the set of all the task under consideration. The net is prepared by examining various relationships that is observed between the tasks recorded in the log. The relationship focused is of causality. The other relationships derived and used are direct succession, parallel, and choice. Causality can be explained as one unique task may be ahead of other unique one in each execution trace/log. Mathematically the relations are deduced as:

• Direct succession: $a > b$; if and only if for some case, a is directly followed by b.

• Causality: $a \rightarrow b$; if and only if for some case, $a > b$ and not $b > a$

• Parallel: $a \| b$; if and only if for some case, $a > b$ and $b > a$

• Choice: $a\#b$; if and only if not $a > b$ and not $b > a$

The 8 steps Alpha Algorithm (α-algorithm) is stated as under (van der Aalst, 2009):

Let W be a workflow log over T. $\alpha(W)$ is defined as follows.

1. $TW = \{t \in T | \exists \sigma \in W t \in \sigma\}$
   TW is a set of all tasks which occur in at least one trace.

2. $TI = \{t \in T | \exists \sigma \in W t = \text{first}(\sigma)\}$
   TI is a set of tasks which occur in trace initially

3. $TO = \{t \in T | \exists \sigma \in W\, t = \text{last}\,(\sigma)\}$
   TO is a set of task which occur in trace terminally

4. $XW = \{(A,B) \,|\, A \subseteq TW \wedge A \neq \varnothing \wedge B \subseteq TW \wedge B \neq \varnothing \wedge \forall a \in A\, \forall b \in B\, a \rightarrow Wb \wedge \forall a_1, a_2 \in A\, a_1 \# Wa_2 \wedge \forall b_1, b_2 \in B\, b_1 \# Wb_2\}$
   XW is a set of all pairs in which places are discovered.

5. $YW = \{(A,B) \in X | \forall (A', B') \in XA \subseteq A' \wedge B \subseteq B' \Rightarrow (A,B) = (A', B')\}$
   YW is a set in which places are identified as pair of set of task for minimal places.

6. $PW = \{p(A,B)|(A,B) \in YW\} \cup \{iw, ow\}$
   PW is the set that contains one place p(A,B) for each pair in YW with input place iw and output place ow.

7. $FW = \{(a, p(A,B))|(A,B) \in YW \wedge a \in A\} \cup \{(p(A,B), b)|(A,B) \in YW \wedge b \in B\} \cup \{(iw,t)|t \in TI\} \cup \{(t, ow)|t \in TO\}$
   FW is the set defining the flow relation.

8. $\alpha(W) = (PW, TW, FW)$.

Van der Aalst et al. (2004b); van Der Aalst (2016)

• Alpha Algorithms Facts

The α-algorithm has some main disadvantages as following:
1. Files of activities in real life are requiring more advanced methods.
2. Some issues such as noisy data files are not handled by it.
3. Can not be accepted as as ideal solution.

However, the algorithm reveals:
1. Basic process mining ideas and concepts in 8 steps.
2. Theoretical limits of process mining.

3.6   BIG DATA (BD) AND HEALTHCARE SYSTEM

The meaning of data refers to properties and features of the organization's elements where in hospital data elements refer to the age, gender, diagnose etc. of patients, due to this fact there are big shift from data orientation to process orientation, while the process means the method of doing the tasks and giving orders of each task using the available resources of any organization. However, data and process have a lot in common. Both are part of business intelligence and the analysis methods of large volumes of data (big data) in order to achieve greater insights. Applying specific algorithms in data and processes is called mining that uncovers hidden patterns and relationships. The goal of data and process mining is to provide users with better decisions. Big data analytic has helped healthcare to be improved by providing personalized medicine and prescriptive analytic, clinical risk intervention and predictive analytic, waste, and care variability reduction, automated external and internal reporting of patient data, standardized medical terms and patient registries and fragmented point solutions. Some areas of improvement are more inspirational than implemented. The level of data generated within healthcare systems is not trivial. With the added adoption of eHealth and wearable technologies the volume of data will continue to increase. This includes electronic health record data, imaging data, patient generated data, sensor data, and other forms of difficult to process data. There is now an even greater need for such environments to pay greater attention to data and information quality. Big data very often means "dirty data" and the fraction of data inaccuracies increases with data volume growth. Human inspection at the big data scale is impossible and there is a desperate need in health service for intelligent tools for accuracy and believably control and handling of information missed. While extensive information in healthcare is now electronic, it fits under the big data umbrella as most is unstructured and difficult to use (Lee et al., 2014) and this information I published in Zayoud et al. (2019a).

● **Characteristics of BD**

BD is expressed with the below characteristics (Lee et al., 2014):
1. **Volume**:
   The amount of generated and stored data. The size of data specifies the value and potential insight- and whether it can be considered big data or not.
2. **Variety**
   The type and nature of data. This helps people who analyze it to effectively use the resulting insight.
3. **Velocity**
   In this meaning, the speed at which the data is generated and processed to meet the demands and challenges that lie in the path of growth and development.
4. **Veracity**
   The data quality of captured data can vary greatly, affecting the accurate analysis (Lee et al., 2014).

However, data need to be analyzed with advanced tools to reveal useful information. In addition, to manage data correctly, it must consider both visible and invisible issues with various components. Information generation algorithms must detect and address many invisible issues that related to the studied domain (Wu et al., 2015a).

● **Big Data Analysis Methods**

Analyzing big data means to answer the following:
1. What are the various challenges in BD that are proposed/confronted by organizations?
2. Or the multi analysis methods of BD proposed/employed to overcome its challenges.

"Observing and understanding the previous trends and extant patterns/themes in the Big Data Analysis (BDA) in research area, evaluating contributions, summarizing

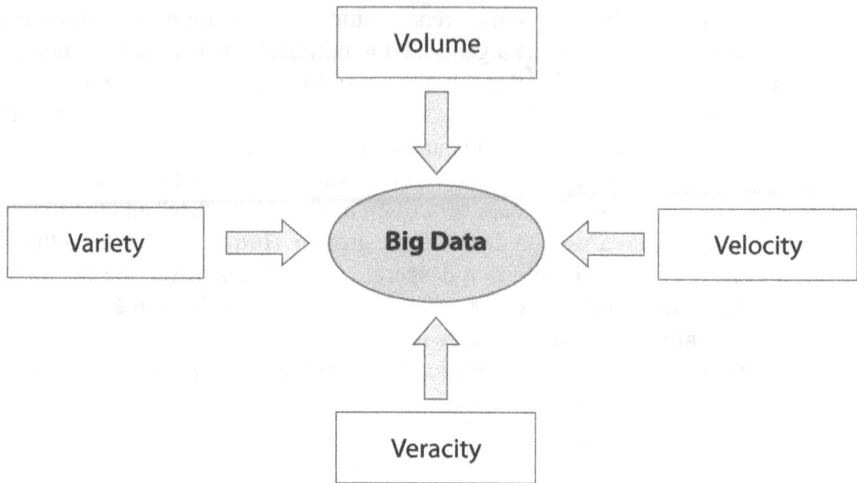

FIGURE 3.10
Characteristics of Big Data

knowledge, identifying limitations, implications, and potential further research avenues to support the academic community in exploring research themes/patterns, needs a systematic literature review (SLR). Thus, to trace the implementation of BD strategies, a profiling method is employed to analyze and identify different Big Data Analysis methods" (Sivarajah et al., 2017). "The potential value of big data is unlocked only when leveraged to drive decision making. To enable such evidence-based decision making, organizations need efficient processes to turn high volumes of fast-moving and diverse data into meaningful insights. The overall process of extracting insights from big data can be broken down into five stages (Labrinidis and Jagadish, 2012). These five stages form the two main sub-processes: data management and analytics. Data management involves processes and supporting technologies to acquire and store data and to prepare and retrieve it for analysis. On the other hand, analytics refers to techniques used to analyze and acquire intelligence from big data. Thus, big data analytics can be viewed as a sub-process in the overall process of "insight extraction" from big data" (Gandomi and Haider, 2015a).

The subsections of this part are reviewing the analytical procedures of (BD) for classified and unclassified data as is presented in (Labrinidis and Jagadish, 2012).

• **Text Analytics**

"Text analytics (text mining) refers to techniques that extract information from textual data. Social network feeds, emails, blogs, online forums, survey responses, corporate documents, news, and call center logs are examples of textual data held by organizations. Text analytics involve statistical analysis, computational linguistics, and machine learning. Text analytics enable businesses to convert large volumes of human generated text into meaningful summaries, which support evidence-based decision-making. For instance, text analytics can be used to predict stock market based on information extracted from financial news. Information extraction (IE) techniques extract structured data from unstructured text. For example, IE algorithms can extract structured information such as drug name, dosage, and frequency from medical prescriptions. Two sub-tasks in IE are Entity Recognition (ER) and Relation Extraction (RE) (Jiang, 2012). ER finds names in text and classifies them into predefined categories such as person, date, location, and organization. RE finds and extracts semantic relationships between entities (e.g., persons, organizations, drugs, genes, etc.) in the text" (Chung, 2014).

• Audio Analytics

"Audio analytics analyze and extract information from unstructured audio data. When applied to human spoken language, audio analytics is also referred to as speech analytics. Since these techniques have mostly been applied to spoken audio, the terms audio analytics and speech analytics are often used interchangeably. Currently, customer call centers and healthcare are the primary application areas of audio analytics. Call centers use audio analytics for efficient analysis of thousands or even millions of hours of recorded calls. These techniques help improve customer experience, evaluate agent performance, enhance sales turnover rates, monitor compliance with different policies (e.g., privacy and security policies), gain insight into customer behavior, and identify product or service issues, among many other tasks. Audio analytics systems can be designed to analyze a live call, formulate cross/up-selling recommendations based on the customer's past and present interactions, and provide feedback to agents in real time. In addition, automated call centers use the Interactive Voice Response (IVR) platforms to identify and handle frustrated callers. In healthcare, audio analytics support diagnosis and treatment of certain medical conditions that affect the patient's communication patterns (e.g., depression, schizophrenia, and cancer) (Hirschberg et al., 2010). Also, audio analytics can help analyze an infant's cries, which contain information about the infant's health and emotional status. The vast amount of data recorded through speech-driven clinical documentation systems is another driver for the adoption of audio analytics in healthcare. Speech analytics follows two common technological approaches: the transcript-based approach (widely known as large-vocabulary continuous speech recognition, LVCSR) and the phonetic-based approach" (Patil, 2010).

• Video Analytics

"Video analytics, also known as video content analysis (VCA), involves a variety of techniques to monitor, analyze, and extract meaningful information from video streams. Although video analytics is still in its infancy compared to other types of data mining various techniques have already been developed for processing real-time as well as pre-recorded videos. The increasing prevalence of closed-circuit television (CCTV) cameras and the booming popularity of video sharing websites are the two leading contributors to the growth of computerized video analysis. A key challenge, however, is the sheer size of video data. To put this into perspective, one second of a high-definition video, in terms of size, is equivalent to over 2000 pages of text (Chui et al., 2011). Now consider that 100 hours of video are uploaded to YouTube every minute (YouTube Statistics, n.d.). Big data technologies turn this challenge into opportunity. Obviating the need for cost-intensive and risk-prone manual processing, big data technologies can be leveraged to automatically sift through and draw intelligence from thousands of hours of video. As a result, the big data technology is the third factor that has contributed to the development of video analytics. The primary application of video analytics in recent years has been in automated security and surveillance systems. In addition to their high cost, labor-based surveillance systems tend to be less effective than automatic systems (e.g., Hakeem et al. (2012) report that security personnel cannot remain focused on surveillance tasks for more than 20 minutes). Video analytics can efficiently and effectively perform surveillance functions such as detecting breaches of restricted zones, identifying objects removed or left unattended, detecting loitering in a specific area, recognizing suspicious activities, and detecting camera tampering, to name a few. Upon detection of a threat, the surveillance system may notify security personnel in real time or trigger an automatic action (e.g., sound alarm, lock doors, or turn on lights). The data generated by CCTV cameras in retail outlets can be extracted for business intelligence. Marketing and operations management are the primary application areas. For instance, smart algorithms can collect demographic information about customers, such as age, gender, and ethnicity. Similarly, retailers can count the number of customers, measure the time they stay in the store, detect their movement patterns, measure their dwell time in different areas, and monitor queues in real time. Valuable insights can be obtained by correlating this information with customer demographics to drive decisions for product placement, price, assortment optimization, promotion design, cross-selling, layout optimization, and staffing" (Abraham and Das, 2010).

• Predictive Analytics

"Predictive analytics comprise a variety of techniques that predict future outcomes based on historical and current data. In practice, predictive analytics can be applied to almost all disciplines from predicting the failure of jet engines based on the stream of data from several thousand sensors, to predicting customers' next moves based on what they buy, when they buy, and even what they say on social media. At its core, predictive analytics seek to uncover patterns and capture relationships in data. Predictive analytics techniques are subdivided into two groups. Some techniques, such as moving averages, attempt to discover the historical patterns in the outcome variables and predict them to the future. Others, such as linear regression, aim to capture the interdependencies between outcome variables and explanatory variables, and exploit them to make predictions. Based on the underlying methodology, techniques can also be categorized into two groups: regression techniques (e.g., multinomial models) and machine learning techniques (e.g., neural networks). Another classification is based on the type of outcome variables: techniques such as linear regression address continuous outcome variables (e.g., sale price of houses), while others such as Random Forests are applied to discrete outcome. Predictive analytics techniques are primarily based on statistical methods. Several factors call for developing new statistical methods for big data. First, conventional statistical methods are rooted in statistical significance: a small sample is obtained from the population and the result is compared with chance to examine the significance of a relationship. The conclusion is then generalized to the entire population. In contrast, big data samples are massive and represent the majority of, if not the entire, population. As a result, the notion of statistical significance is not that relevant to big data. Secondly, in terms of computational efficiency, many conventional methods for small samples do not scale up to big data. The third factor corresponds to the distinctive features inherent in big data: heterogeneity, noise accumulation, spurious correlations, and incidental endogeneity" (Fan et al., 2014).

They are described as following:

- "Heterogeneity. Big data are often obtained from different sources and represent information from different sub-populations. As a result, big data are highly heterogeneous. The sub-population data in small samples are deemed outliers because of their insufficient frequency. However, the sheer size of big data sets creates the unique opportunity to model the heterogeneity arising from sub-population data, which would require sophisticated statistical techniques.

- Noise accumulation. Estimating predictive models for big data often involves the simultaneous estimation of several parameters. The accumulated estimation error (or noise) for different parameters could dominate the magnitudes of variables that have true effects within the model. In other words, some variables with significant explanatory power might be overlooked as a result of noise accumulation.

- Spurious correlation. For BD, spurious correlation refers to uncorrelated variables being falsely found to be correlated due to the massive size of the dataset" (Fan and Lv, 2008) expresses this characteristic based on a simulation example, where the coefficient of the correlations among non dependent default variables is studied to maximize the size of a dataset. In practical situation and when big dimensional results appear, it shows how those un related and non dependent variables are erroneously proven to be correlated.

- Incidental endogeneity. A common assumption in regression analysis is the exogeneity assumption: the explanatory variables, or predictors, are independent of the residual term. The validity of most statistical methods used in regression analysis depends on this assumption. In other words, the existence of incidental endogeneity (i.e., the dependence of the residual term on some of the predictors) undermines the validity of the statistical methods used for regression analysis. Although the exogeneity assumption is usually met in small samples, incidental endogeneity is commonly present in big data. It is worthwhile to mention that, in contrast to spurious correlation, incidental endogeneity refers to a genuine relationship between variables and the error term. The irrelevance of statistical significance, the challenges of computational efficiency, and the unique characteristics of big data discussed above highlight the need to develop new statistical techniques to gain insights from predictive models (Fan and Lv, 2008).

• Challenges

"The advantages of implemeting BD are factual and substantial, but it maintains some of challenges that must be highlighted to fully realize the potential of BD. However, big data has some challenges such as the function of the characteristics of BD, or by its existing analysis methods and models, and some, through the limitations of current data processing system" (Sivarajah et al., 2017). "The research studies regarding BD challenges have paid attention to the difficulties of understanding the meaning of big data, and the decision-making of what data are generated. In addition, how it is collected and their issues of privacy and ethical considerations relevant to mining such data or processes. In Tole et al. (2013), it asserts that building a viable solution for large and multifaceted data is a challenge that businesses are constantly learning and then implementing new approaches. For example, one the biggest problems regarding BD is the infrastructure's high costs. Hardware equipment is very expensive even with the availability of cloud computing technologies. Furthermore, to sort through data, so that valuable information can be constructed, human analysis is often required. While the computing technologies required to facilitate these data are keeping pace, the human expertise and talents business leaders require to leverage BD are lagging, this proves to be another big challenge. As reported by Akerkar (2014) and Zicari (2014), the broad challenges of BD can be grouped into three main categories, based on the data life cycle: data, process, and management challenges:

- *Data challenges* relate to the characteristics of data itself as mentioned in the previous section.
- *Process challenges* are related to series of how techniques: how to capture data, how to integrate data, how to transform data, how to select the right model for analysis and how to provide the results.
- *Management challenges* cover for example privacy, security, governance, and ethical aspects" (Sivarajah et al., 2017).

Figure 3.11 shows the big data challenges – as adapted from Akerkar (2014) and Zicari (2014).

• Process Mining (PM) and Big Data in Healthcare

"Health services sector has been growing largely over the last decade and healthcare services became more complex and costly, amplified by a poor healthcare delivery system. Healthcare is a highly interconnected dynamic environment where individuals and teams contribute in order to serve patients' demand. The focus of this study is to improve the processes in the services sectors like the medical communities. This can be achieved by studying the current processes and improving them or creating new methodologies in order to improve the health care systems as all. Management consultant HSPI (an Italian management consulting company) recently published an overview of 113 process mining projects (Gonella, 2016). Figure 3.12 shows a breakdown of projects categorized by industry. Process mining is mostly used by service-oriented organizations (including healthcare) and not so much in the manufacturing industry. This observation is also seen in other studies described in the literature.

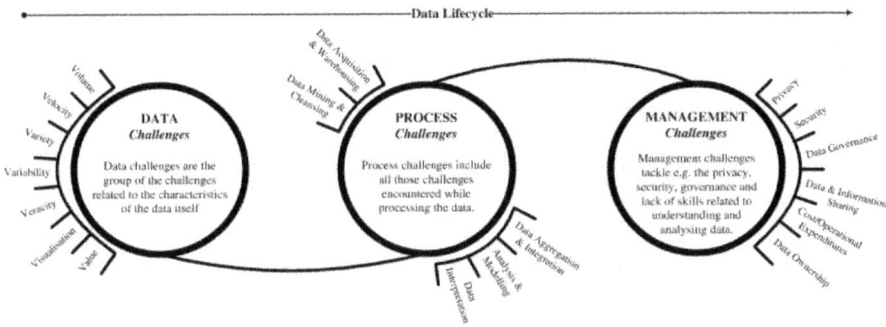

FIGURE 3.11
Classification of Big Data challenges
Source: Akerkar (2014); Zicari (2014).

A reason can be that data produced in the service industry is of higher quality and granularity. It is also possible that processes in service industry have a relatively short lead time, where construction and maintenance processes take longer. A large timeframe is than needed to ensure having enough complete processes to mine. Why process mining is less popular in the manufacturing industry is remarkable. Mass production, for example, is characterized by its strict process models. Perhaps this industry did not discover the potential of process mining yet. Figure 3.12 shows the demand of healthcare sector on process mining which makes it an attractive field to implement the relevant algorithms and tools or develop new ones" Gonella (2016).

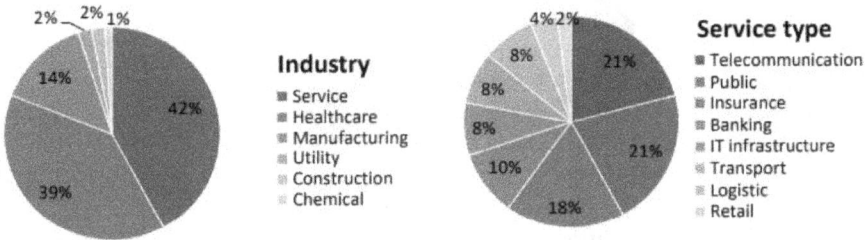

FIGURE 3.12
The 113 process mining projects as categorized by industry
Source: Chui et al. (2007).

3.7   PROBLEMS DEFINITIONS

The first area to focus on, to develop an efficient and effective healthcare system is to have real data imported by efficient tools that analyzes the data, convert it into event logs and discover the current process. Discovering the process leads us to have the knowledge of the weaknesses and strengths and provides guidelines of how to improve the process or innovate a new one; thus, leading to a more effective and efficient structure. Simulation designs are effective ways to minimize difficulties especially for prediction manners that referred to policies and strategies. Moreover, simulations are procedures that help taking important decisions due to the experimental nature of their results that provide management of almost real situations and implementation. Busy and complex healthcare systems provide big challenges to managers and decision makers who should be able to serve the high demands constrained by limited budget and high costs of healthcare services. The highest number of patients should be handled within a limited period in order to ensure patient satisfaction (reduce waiting times) and increase hospital's revenue (reduce cost). The most common weaknesses in any health care service provider are: Timeliness of Care, Quality of Care, Errors in Care Delivery, Complexity of healthcare system, and Cost effectiveness of care (Akshay, 2012).

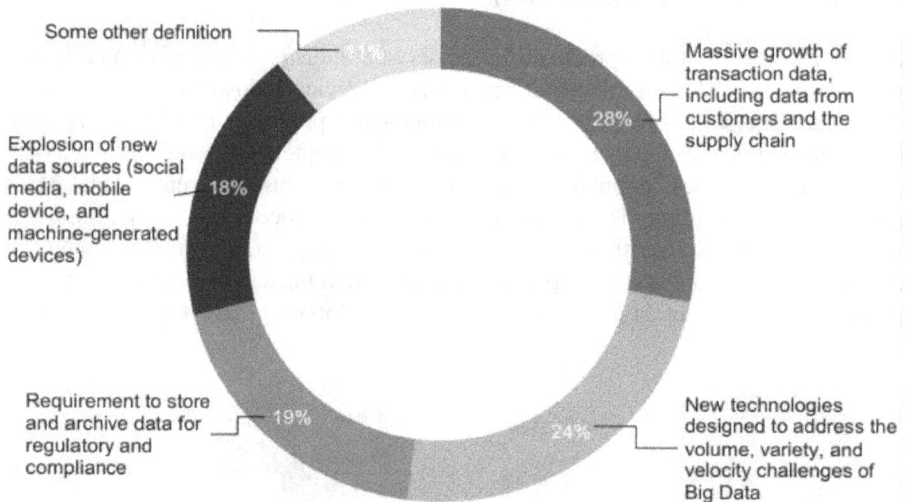

FIGURE 3.13
An online survey of definitions of big data for 154 global executives in April 2012
Source: Chung (2014) and Jiang (2012).

Moreover, data sets grow rapidly – The world's technological per capacity to store information has roughly doubled every 40 months since the 1980s (MIKE2.0, 2013); as of 2012, every day 2.5 exabytes (2.5×1018) of data are generated (Chung, 2014) By 2025, IDC predicts there will be 163 zettabytes of data. One question for large enterprises is determining who should own big-data initiatives that affect the entire organization (Data, 2010). Relational database management systems and desktop statistics- and visualization-packages often have difficulty handling big data. The work may require massively parallel software running on tens, hundreds, or even thousands of servers. What counts as BD varies depending on the capabilities of the users and their tools, and expanding capabilities make big data a moving target. For some organizations, facing hundreds of gigabytes of data for the first time may trigger a need to reconsider data management options. For others, it may take tens or hundreds of terabytes before data size becomes a significant consideration, as presented in Appendix B the limitation of SQL session to handle huge amount of date as I wrote in Zayoud et al. (2019b).

3.8 CONCLUSION

As a summary to Chapter 3, process mining methods are defined and studied. The α algorithm is explained in detail, all the algorithm steps and facts are mentioned. Issues and challenges of using mining techniques for processes are explained as well, BD is the biggest challenge to organizations especially HC system is also presented.

Where healthcare is a highly interconnected dynamic environment where individuals and teams contribute to serve patients' demand. Because my focus of this book, is to improve the processes in the services sectors like the medical communities. This can be achieved by studying the current processes and improving them or creating new methodologies to improve the health care systems as all. Hence, accurate log files are required and a reliable method that is able to handle huge amount of data that may considered as big data is also a must to produce better process model that has a great impact on the organization in many aspects, in our situation which the HC system is studied, there are many things should be handled such is patients' satisfactions, management satisfactions and health improvement in general. As seen earlier in this chapter, the current methods have issues in handling the increasing demands of the organizations and the huge amount of data that are generated during years. Hence, I have studied all those issues and based on the α algorithm, I have introduced a new framework to overcome many challenges as well be seen in the next chapter of this book, and proposes the extended α algorithm and named as the β algorithm, and proving this new algorithm by showing the statistical and probabilistic results in addition to mathematical calculations and simulations to validate the new algorithm by extracting a required knowledge to improve healthcare system.

4 The Proposed β Algorithm for Healthcare System

4.1 INTRODUCTION

Process Learning (der Aalst, 2011b) is an important methodology for organizations to enhance their workflows and processes as well that they deploy especially after the huge increase of the amount of data those organizations are dealing with on daily bases (Smola and Schölkopf, 2004). Process learning is defined as an implementation of the practice that is able to find a clear understanding of atomic events that happen in some domains in addition to the correlation and dependency between those events (der Aalst, 2011b). It is known in the literature that there are a few algorithms which can deal with the problem of process learning (Aha et al., 1991a; Dumais et al., 1998; Schölkopf et al., 1999a; Smola and Schölkopf, 2004; van der Aalst et al., 2006a; Van der Aalst and Weijters, 2004a; Weijters et al., 2006c). However, each of them has its deficiency as seen in Section 5.3. The probabilistic fact of dependencies between events is presented, and a new proposed learning algorithm is defined in this chapter.

"Process learning is the knowledge discovery, a part of data mining and process mining. The process of analysing different types of data is called data mining. Process mining is the mean of discovering the **process** after **analysing** different **sets of data** that are provided and stored in a management information system of a particular organization. Both, data and process mining, aim to maintain business entities and improve performance. The identity of a process is specified depending on data types and application domains. Having mentioned this, healthcare data generates knowledge that discovers a healthcare process. This knowledge includes events that identify the nature of this field like patient arrival, treatments, diagnoses, etc. (Kassirer et al., 1991). Process learning is the focus of this chapter. Because the process mining is considered a hot topic since it is driven by industrial and organizational needs to enhance processes" (Kassirer et al., 1991). The following are considered classifications of process mining:

- Detecting Process,
- Ensuring Congruity, and
- Development.

4.2 PROCESS MINING CLASSIFICATIONS

Process discovery is a knowledge extraction process that is based on event logs (Conforti et al., 2017). "The outcome of the mining process is an accurate process model.

DOI: 10.1201/9781003366577-4

This is achieved through processing event log files after ensuring they are coherent with the actual events in a given system." The extracted model can be enhanced by monitoring system procedures, studying their limitations and overcoming overall performance deficiencies. This can be achieved by improving some events or other factors in the given event logs.

Process mining can broadly be classified into the following major categories:
- Local algorithms: α and Heuristic.
- Global algorithms: Genetic and fuzzy.

Local algorithms are based on local information about events in a log, that is, used to set the dependencies between these events by extracting information about what tasks directly precede or directly follow each other in that log.

The global algorithms or non-local non-free-choice constructs, however, cannot be determined by only looking at the direct successors and predecessors which is the local context of a task in a log. However, all dependency relations are inferred based on local information in that log. Moreover, most techniques cannot mine non-local non-free-choice because they are based on local information in those logs.

The techniques that do not mine local non-free-choice cannot do so because their representations do not support such constructs. On the other hand, the non-free choice constructs combine synchronization and choice (Alves de Medeiros, 2006).

Both local and global algorithms have their own deficiencies, limitations, and challenges. The choice of which algorithm to choose and build the next step on, depends on what needs to be done. The difficulties are uncountable and not limited such as the following:
- Input data that is incomplete or noisy.
- Distinguishing simultaneous sequences, and forks.
- Achieving random repetitions and classified sets.
- Detecting events in repetition mode from the events of a loop.
- Handling starting and exiting points in fuzzy processes.
- Detecting various types of processes with their classifiers.

The following table is focusing on various difficulties of my new approach of this book versus some other famous algorithms that are mentioned in literature.

Where $\sqrt{}$: means that an algorithm can handle the mentioned challenge. $\times$: means can't be handled by this algorithm (Zayoud et al., 2019a).

By comparing the available most common approaches, many of the limitations can be resolved by applying a learning method that builds knowledge accumulatively by having different data sets, and events logs to ensure accurate model as a result of learning the process. From the challenges mentioned above, a good learning approach would identify the following as presented in my work (Zayoud et al., 2019a):
- Events that are parallel: begin and ends at the exact time.
- Events that are ored: when at least of of the ored events happened.

TABLE 4.1

The Challenges and Difficulties of Process Mining Algorithms Responds

Item	β	α	Fuzzy	Genetic	Heuristic
Noisy input data:	✓	×	✓	✓	✓
Concurrency:	✓	✓	✓	✓	✓
Loops:	✓	×	×	✓	×
Repeated Squences:	✓	×	×	✓	×
Fuzzy process:	✓	×	✓	×	✓
Variants detection:	×	×	✓	×	×

- Events that are xored: when two events are xored that is mean they can never happen at the same time.
- implication: this relation means one event should be done to let the next event to start.
- repeated patterns:means some events hapened in the same order of each process.
- loops: the whole events in a process are repeated in the sam eorder.

As seen in chapter three of this book, the famous and known algorithms that represent business "process mining algorithms" (Smola and Schölkopf, 2004) are: α algorithm (Duda et al., 2012), genetic algorithm (De Medeiros and Weijters, 2005), heuristics miner (Weijters et al., 2006d), and region miner (van Dongen et al., 2007b). This chapter is prepared as follows: Section 4.3 presents mainly process mining algorithms applications. Section 4.4 presents the challenges that face basic process mining algorithms. Section 4.5 demonstrates a comparison between those algorithms with the innovative algorithm in this chapter. Section 4.6 clarifies the established platform of "probabilistic learning" and explains the conversion method of the learned process to predicate logic. In Section 4.8, the application of that platform that is designed for healthcare. Section 4.9 talks about Bayesian networks, Section 4.10 explains Petri Nets, and Sections 4.11 and 4.12 mention the tools and method of data management. A comparison between the proposed solution with the existing process learning algorithms that have been proposed in literature. Finally Section 4.13 concludes this chapter with some extension.

4.3 APPLICATIONS OF PROCESS MINING ALGORITHMS

A quick overview of the famous algorithms and their applications of process mining, is provided in this section.

The innovative (IS) is designed in a distributed manner, they also include logs of events that are also distributed among multi physical machines which is also implemented in what is called Map-Reduce (Evermann and Assadipour, 2014), knowing that the "α algorithm" is developed to approach a "Map-Reduce design" based on the scalability concept of the design of Map-Reduce. This design implementation in terms of its performance and scalability is tested while experiencing real-life applications. Moreover, when log files have knowledge about the process instance, such that an action is also called an "instance" task that has a timestamp, hence in this situation, when there is a knowledge about instance tasks, the "α" algorithm will re detects a correct workflow net of a log with the condition of log free of noise. In contrast, when there are many problems that cannot be resolved by only α, hence "Map-Reduce maybe a solution in such situation because it is a programming approach for large scale data processing in a distributed computing environment, and it works according to important functions such as the map function accepts a series of pairs from an input reader and provides a series of pairs as an output, and then a shuffle function, which sorts and collects the values for different keys, providing output as a series of tuples, then the reduce function takes this input and provides the output as series of key pairs. The important aspect in implementing a Map-Reduce based algorithm is the combination of suitable sequences of map and reduce functions with appropriate data types for keys and values. To compute the log-based ordering relations for the α algorithm, it is required two sets of mappers and reducers, following each other" (Zayoud et al., 2019a).

Fuzzy miner algorithm (Günther and Van Der Aalst, 2007a) depends on a different approach which provides an actice view of a process by the method of generalization that is defined as representation of specific details that can be changed over time. The concept of fuzzy miner is about the existance of more than one matrix as an example, one matrix is called the "significance" and the other one is "correlation", which are factors can be calculated. They begin by calculating the repeated appearance of an event, and its place in a log file, if this event is repeated frequently, hence a "precedence relation" is detected. Then, the "correlation" metric provides information about those events and if they are related significantly. However, those measurements are based on the shared data between them, in addition to the similarity among activities names. Those factors about events in terms of their strong correlation and significance as well are the building blocks of generating what is called a "process model". Knowing that the strong behavior of significance behavior is detected, if the behavior is strongly correlated. In the case of the weak significance events, hence they have to join clusters, or left out if the behavior shows weak correlation as well. As a result, views of events log are established continuedly using the detected measures of "correlation" and "significance", those views help analysts to zoom in and out on particular aspect of a process model which can open the way to detect issues and have solutions for them. As an example of the previous method, it was implemented "for an application of process mining for internal transaction fraud mitigation in a famous company, such that the process diagnostics only concerns on a global view of the business process, in order to help the analysts and domain experts to reveal weaknesses and problems in the business process and discovers any fraud cases may happen and exposes opportunities to commit fraud in the followed process" (Jans et al., 2011). Since, it is given real life example, hence I need to highlight the importance of deploying methods of process mining on real life applications, and as we all know and can see especially in the current situation of all over the world, "healthcare" domain is a dominant field and can affect all life aspects especially the economic factors. This domain is rich environments in terms of information, but with few process models, and shortage of powerful analysis tools to recognize unseen relations or data trends especially if it is considered (BG) (Soni et al., 2011). Healthcare system are very attractive to be studied and analyzed using process mining methods because the collected data in this domain has uncounted issues and mistakes that belong to different categories even though it is provided by expertise in this domain. As examples of these problems the wrong diagnosis and treatment, in addition to unnecessary tests for diagnoses are taken by patients that may not add any contribution towards effective diagnosis of a disease, which means as a result the lack of accuracy. However, problems in this domain are not only related to wrong diagnosis, they are extended to variety of factors that can cause catastrophic. After this explanation I chose healthcare systems to be analyzed and studied using process mining methods and they are deployed to test the validity of the proposed method in this book. In reference to this discussion "Genetic algorithm" is suitable to detect the features that have a great impact on minimizing the tests that are usually required of any patient for providing more accurate diagnosis especially in reference to heart ailments as an example (Anbarasi et al., 2010). In the same example of heart problems, the number of characteristics are reduced from "13" to "6" using genetic search method. Moreover, they became three characteristics that help to expect the accurate diagnosis of a patient using the techniques of "Genetic" algorithm (Anbarasi et al., 2010).

As I mentioned previously in this book that the impact of healthcare systems is huge, and the growing needs of this domain, especially on hospitals that are the building blocks of these systems, motivate health systems providers to seek enhancement of their care processes and develop efficiency in addition to guaranteeing the quality of the provided care services (Kaymak et al., 2012). Seeking improvement in this domain is not an easy task because process modelling is a critical phase, starting from providing a behavior of the process and ending with optimizing and utilizing that process. Knowing that, some previous applications of process mining on healthcare systems produced complex models. In Kaymak et al. (2012) can be seen the weaknesses of those process mining methods in terms of achieving accepted process models, especially if there are "well-defined" clinical processes. Achieving complex process models in this domain is considered as a challenge of using the methods of process mining, "Heuristic miner" algorithm is started to take its place in this domain and provide some valuable solutions to many of the difficulties and limitations using the basic discovery of the model that defines the process. This concept made this algorithm can identify the most valuable activities, but after it is been tested by the experts of medical domain, they came up with a conclusion that there are some sequences in the identified process model which do not add any value in terms of medical perspectives, and the output model cannot describe some relationships that usually should be there between events such as causality relations. As an example, "an increase in the respiratory rate seems to trigger an increase of the carbon dioxide level in the blood". In Kaymak et al. (2012) provides some recommendations such as improving the quality of data sets that may affect positively the resultant model, these new rules on data to concentrate only on a part of events not on all of them directs the heuristic algorithm to the right position in gaining knowledge, and the output model of a process should be more fit and suitable for events belong to the domain of study as presented and published in my article (Zayoud et al., 2019a).

4.4 PROCESS MINING CHALLENGES

As seen before, the process mining is the practice of achieving knowledge based on the available event logs in a specific organization (der Aalst, 2011b). Process mining techniques develop, and discover processes through the merge of data mining techniques and business process management. Data mining deals with large data sets whereas business process management focuses on modeling those sets. Having said that, process mining is considered an intermediate phase between the two, it combines data analysis with modeling. Moreover, process mining can handle raw data or event logs of any organization. Since data mining is a method that discovers, analyzes and detects data, it has no direct link with the business processes. In the other hand, process mining discovers, controls, and improves actual business processes based on data acquired from information systems equipped in organizations. By analyzing data derived from those equipped systems that support processes, process mining gives a true, end-to-end view of how business processes operate (Sani et al., 2018). Data mining analyses static information which is data that is available at the time of analysis. Process mining on the other hand looks at how the data was actually created. Process mining techniques also allow users to generate processes dynamically based on the most recent data. Process mining can even provide a real-time view of business processes through a live feed (Sani et al., 2018). Process mining is the extraction of models of business processes, from log files maintained by organizations, to enable business decision makers to better understand and optimize business activities.

The outcome of process mining is the process model that describes this process. This model can be graphical, for analysts and management use, or formal, for mathematical analysis use. As mentioned earlier, there are many algorithms that target process mining (van der Aalst et al., 2006b). Choosing which algorithm or method to implement, depends on the facts of each one and on the field of study. Knowing the best algorithm is a challenge for businesses that need to deploy process mining, and for researchers to evaluate new developments.

Choosing a suitable process mining algorithm follows the following milestones:
- Analyzing the log.
- Identifying all process instances.
- Defining some relations among events.

Once all event relations are defined, it is possible to combine them in order to construct the mined model. However, even if several approaches are available, many problems are still unsolved. In Van der Aalst and Weijters (2004a), some of those problems are identified. The following is a summary of those problems:

- ▶ Some process models may have the same event appearing several times, in different positions.
- ▶ Many times, logs report a lot of data not used by mining algorithms. For example, detailed timing information, such as distinguishing the duration time of specific event which can be used to guarantee the accuracy of mined models.
- ▶ Current mining algorithms do not perform a **holistic mining** of different perspectives, coming from different sources; for example, not only the control-flow, but also a social network with the interactions between the activity originators (creating a global process description).
- ▶ Visualization of mining results: present the results of process mining in a way that people can gain insights in the process.
- ▶ Dealing with noise and incompleteness, **noise** identifies uncommon behavior that should not be described in the mined model.
- ▶ Incompleteness represents the lack of some information required for performing the mining task.

Almost all business logs are affected by the last two mentioned problems, and process mining algorithms are not always able to properly solve them, but applying new extensions on the existed methods to fit better the requirements of a particular domain of business is an effective way to resolve different kind of issues (Zayoud et al., 2019a).

4.5 A DIFFERENTIATION STUDY ON LEARNING PROCESSES APPROACHES

An accepted benchmark is not easy to found for evaluating and comparing the different proposed process mining algorithms which makes it difficult to select a suitable process mining algorithm for a given enterprise or application domain. However, process mining algorithms are used to mine business process models using process logs. The mined models are compared against the current process models of the enterprise for conformance checking that is a basic step of process mining. Next step is to discover more efficient, streamlined business process models. To advantage of one of the methods in process mining over others that have different features, is to produce mined models that are semantically similar and structurally equal to the original models or better than the original models, for an interested. Recent research and software prototypes have attempted to provide such an evaluation framework, e.g., Weijters et al. (2006d) empirically evaluating all available process mining algorithms against the business models provided by a given enterprise is usually computationally expensive and time consuming. Therefore, the main challenge of these process model mining algorithms is the need to be regularly re-evaluated against these changing models for conformance checking, re-engineering or discovery of more streamlined, improved models.

The starting point of comparing between different process mining algorithms, is to present how each one of them is discovering events' relations. For example, in heuristics miner algorithm, the establishment of a graph that represents the dependency between events and their frequencies in terms of metric to clarify if there is a dependency between two events in a log file or not. The problem of this algorithm is far from being completed due to it can not handle short loops,or the logic dependency relations between different events like AND, XOR, Split/Join in addition to the problem of non-observable events (Weijters et al., 2006d). On other hand, genetic algorithm (De Medeiros and Weijters, 2005) is mainly used for mining noisy event logs and it is considered as adaptive search method to mimic the process of evolution. The initial experiences showed that to represent individuals in terms of Petri Net (Murata, 1989a) is not convenient, and the expressive power of Petri Net will be limited in some cases. However, this method use a new representation called casual matrix that shows the direct cause of event by applying routing techniques, for example, when one event is the single cause of another event, there is a sequential routing.This algorithm produces three kinds of routing between events which are sequential, parallel or choice routing. Different outing depends on the relations between different events and which cause the other. The problem of the genetic algorithm is some relations between events are not handle. The region miner method (Weijters et al., 2006d) that establishes a connection between transitions systems and Petri Nets through what is called net synthesis or state-based model that shows which states a process can be in and what are the possible transitions between these states. This model can be transformed into a Petri Net that explicitly shows causality, concurrency and conflicts between transitions. This algorithm seems to be a transition system whereas event logs do not carry state information, The second problem with region algorithm is it assumes the transition system shows all possible transitions between states, while in process mining, the assumption usually is the event logs do not contain all possible sequences of events. The α algorithm is the base of our proposed framework, and it has many extensions. It is designed to handle concurrency in processes, and proven to correctly mine processes where the process can be modelled by a structured workflow net. α algorithm requires a complete event log. Original α algorithm studies the following relations (der Aalst, 2011b):

▶ Fellow

▶ Causality

▶ Parallel

▶ Un related

The above defined sets are not closed forms since important operators are not mentioned like the mutual explosion, and, or relations between events. It is developed to mine concurrent processes and it is proven to mine processes represented by structured nets from noise-free logs that is not practical in real life applications (Zayoud et al., 2019a). A comparison between the "α" and the "β" is shown later of this chapter.

Table 4.2 shows comparison between the proposed algorithm and other common process mining algorithms regarding their handling for relations between events. The comparison above shows the most common solutions and how each one is manging

TABLE 4.2

Events' Relations vs. Process Mining Algorithms

Item	β	α	Fuzzy	Genetic	Heuristic
AND operation: $\wedge$	√	√	×	×	×
OR operation: $\vee$	√	√	×	×	×
XOR operation: $\oplus$	√	×	×	×	×
Implication : $\rightarrow$	√	√	×	×	×
Partial Parallelism : $\leftrightarrow$	√	√	×	×	×
Full Parallelism : $\Leftrightarrow$	√	×	×	×	×
Probabilistic	√	×	×	×	×
Local	√	√	×	×	√
Global	×	×	√	√	×

real applications that have complicated processes and events in these processes and they are not simple enough to be manged by solution that do not take the mentioned logical relations between those events. Therefore, I have proposed β algorithm as extension to α, algorithm that is simple and easy algorithm that discover the workflow of the process but cannot be implemented in real life applications due to its simplicity and lack in managing important logical relations between events, the proposed framework is explained in detail with all mathematical proofs in the next sections of this chapter.

4.6 PROPOSED FRAMEWORK β ALGORITHM

This chapter introduces an innovative approach as a framework for learning log files to detect processes. "This framework is probabilistic and it learns the process in an accumulative manner. The learning process is divided into phases. First we start with learning the set of events in the log files then we calculate the dependencies among those events. After learning the dependencies, we calculate the probability distributions of those events and their correlation. In this section, the phases of this learning framework are described in details in my article" (Zayoud et al., 2019a). We first start with defining and learning the events.

• Event Learning

In this phase, we get input logs an identify events in those log files. Mathematically, we define the set of events as follows:

$$\Lambda = \{\lambda_1, \lambda_2, \ldots, \lambda_n\} \tag{4.1}$$

where n is the maximum number of events. We update the set Λ according to the following equation.

$$\forall x \in \mathbb{L}, \quad x \notin \Lambda \Rightarrow \Lambda = \Lambda \cup \{x\} \tag{4.2}$$

where $\mathbb{L}$ is the set of log files and x is an event such that $x \in \mathbb{L}$. In other words, we only add events to the set of events λ is those events did not exist before in Λ (Zayoud et al., 2019a).

• Dependency Learning

"Dependency matrix is a square matrix that shows the dependencies between events. If $\|\Lambda\| = n$, then the matrix is $n \times n$ dimension. We denote this dependency matrix with $\mathbb{D}$. We start the dependency learning by having all values of $\mathbb{D}$ to be zero then we update $\mathbb{D}$ based on what we learn from log files. Dependency matrix $\mathbb{D}$ represents the correlation between events. If event λ_y depends on event λ_x then $\mathbb{D}(y,x) = 1$ otherwise $\mathbb{D}(y,x) = 0$. The algorithm to learn events from log files is presented in details in Appendix A. The complexity of this algorithm is $O(n^2)$. Solving the problem with an algorithm that has less complexity is possible however it is out of the scope of this book" (Zayoud et al., 2019a).

• Identifications of Events Relationships

Next, the process recognition to detect correlations and classifies them logically, reaching to this phase in gaining knowledge, the process will be able to build a work-flow model of the system. In this phase, a log full of valuable information about the duration of every event is needed to detect correlations among events. The following possible relationship are:

▶ "No relation"
▶ "AND" with symbol $\land$
▶ "OR" with symbol $\lor$
▶ "XOR" with symbol $\oplus$
▶ "Implication" with symbol $\rightarrow$
▶ "Followship" with symbol $\dashrightarrow$
▶ "Partial Parallelism" with symbol $\leftrightarrow$
▶ "Full Parallelism" with symbol $\Leftrightarrow$
▶ "Sequence" with symbol $\mapsto$
▶ "Repeated Sequence" with symbol $\frown$
▶ "Loop" with symbol $\circlearrowleft$

When more than action is under **AND** relation: "λ_i" and "λ_j", "λ_i" $\land$ "λ_j" has a logical value of "1" (IFF) for all occurrences of those events $t_j \le t_i \le t_j + \tau_j$ or $t_i \le t_j \le t_i + \tau_i$. By mathematics:

$$(\lambda_i \land \lambda_j) \Rightarrow \forall t_i, \tau_i \in \lambda_i, \quad \forall t_j, \tau_j \in \lambda_j, \quad (t_j \le t_i \le t_i + \tau_j) \lor (t_i \le t_j \le t_i + \tau_i) \tag{4.3}$$

To clarify the mathematical expression above, it can said that the two actions that are executed simultaneously means the time of one is beginning can be during the time of the second action. For two events "λ_i" and "λ_j" under **OR** relation: "λ_i" $\vee$ "λ_j" is logically equals to "1" (IFF) in the occurrences of the two actions $t_j \leq t_i$ or $t_i \leq t_j$. Mathematically:

$$(\lambda_i \vee \lambda_j) \to \forall t_i \in \lambda_i, \quad \forall t_j \in \lambda_j, \quad \exists (t_j \leq t_i) \vee (t_i \leq t_j) \tag{4.4}$$

It can be said in this relation that the two actions are executed simultaneously or only one of them can be executed.

When there are "λ_i" and "λ_j" as two actions under **XOR** relation : "λ_i" $\oplus$ "λ_j" is logically equals "1" (IFF) for all the occurences of those actions: $t_i < t_j$ or $t_i > t_j + \tau_j$ or $t_j < t_i$ or $t_j > t_i + \tau_i$, mathematically:

$$(\lambda_i \oplus \lambda_j) \Rightarrow \forall t_i, \tau_i \in \lambda_i, \quad \forall t_j, \tau_j \in \lambda_j, \tag{4.5}$$

$$(t_i < t_j \vee t_i > t_j + \tau_j) \vee (t_j < t_i \vee t_j > t_i + \tau_i) \tag{4.6}$$

Hence, It is not allowed to have any common time of their execution.

When "λ_j" **depends** on "λ_i": "λ_i" $\to$ "λ_j" is logically equals to "1" (IFF) each execuation of "λ_i" is started before the execuation of "λ_j". $t_j < t_j + \tau_j$. Mathematically:

$$(\lambda_i \to \lambda_j) \Rightarrow \forall t_i, \tau_i \in \lambda_i, \quad \forall t_j \in \lambda_j, \quad (t_j \geq t_i + \tau_i) \tag{4.7}$$

As a result, the execuation time of one action is after the execuation of the other one. The time is a key factor to identify which one is the follower.

If "λ_j" **follows** "λ_i": "λ_i" $--\!\!\to$ "λ_j" has a logic value of "1" (IFF) the execuation time of "λ_i" comes before the time "λ_j". According to mathematics expressions:

$$(\lambda_i --\!\!\to \lambda_j) \Rightarrow \exists t_i, \tau_i \in \lambda_i, \quad \exists t_j \in \lambda_j, \quad (t_j \geq t_i + \tau_i) \tag{4.8}$$

In the above expression, the follow means that the execuation time of an action somtimes the occurrence of another.

Talking about "Parallism" relation is good to know that is either one of the two types "Full" or "Partial". If it is of type **Partial**, then the actions that are categorized under this type (IFF) one action begins its execution together while their ending time might be not the same and their implementation's durations are different. It is written below as mathematical expression:

$$(\lambda_i \leftrightarrow \lambda_j) \Rightarrow \forall t_i \in \lambda_i, \quad \forall t_j \in \lambda_j, \quad (t_j = t_i) \tag{4.9}$$

The second type which is the **Full** is recognized when the two actions under this category have the same execution's durations and they begin and end at same time. The next is a mathematical expression of this type:

$$(\lambda_i \Leftrightarrow \lambda_j) \Rightarrow \forall t_i, \tau_i \in \lambda_i, \quad \forall t_j, \tau_j \in \lambda_j, \quad (t_j = t_i) \wedge (\tau_i = \tau_j) \tag{4.10}$$

When many actions in a log are classified with "dependency" relation in a sequential manner, hence those actions are belong to a **Sequence**. Having "λ_i, λ_j", and "λ_k", those activities are a "Sequence" (IFF) "$\lambda_i \to \lambda_j$" and $\lambda_j \to \lambda_k$. The greatest achieved length of any sequence can be assumed as the symbol $\|\Lambda\|$. The multiple occurance of any sequence is not allowed and the concept of "cycles" as well, hence any action λ in any sequence is occured once only. Identifying a "Sequence" is simple task and presented as the following mathematical expression:

$$(\lambda_i \mapsto \lambda_j) \Rightarrow \forall \lambda_i, \lambda_j, (\lambda_i \to \lambda_j) \vee \exists \lambda_k \| \lambda_i \to \lambda_k \wedge \lambda_k \to \lambda_j \quad (4.11)$$

A normal "Sequence" is considered as **"Repeated Sequence"** if its actions that belong to it is occurred more than once in a log file with a condition of having in between any two occurrence other actions that not belong to that "Sequence". This correlation is described in Mathematics as the following expression:

$$(\lambda_i \mapsto \lambda_j)^\frown \Rightarrow \|(\lambda_i \mapsto \lambda_j)\| \geq 2 \wedge \exists \lambda_k \notin (\lambda_i \mapsto \lambda_j) \| t_i(\lambda_i \mapsto \lambda_j) \leq t_k \leq t_j(\lambda_i \mapsto \lambda_j)$$
$$(4.12)$$

If any "Sequence" is executed more than once in a log file and those two or more occurrences are not containing any other action that not belong to that "Sequence", hence the resultant case is considered a **"Loop"**. The below expression is the mathematical format of the "Loop":

$$(\lambda_i \mapsto \lambda_j)^\circlearrowleft \Rightarrow \|(\lambda_i \mapsto \lambda_j)\| \geq 2 \wedge \nexists \lambda_k \notin (\lambda_i \mapsto \lambda_j) \| t_i(\lambda_i \mapsto \lambda_j) \leq t_k \leq t_j(\lambda_i \mapsto \lambda_j)$$
$$(4.13)$$

The "Repeated Sequence" is the base to have a "Loop" knowing that no other actions outside that Sequence will separate the frequent occurrence of that sequence (Zayoud et al., 2019a).

• Event Recognition Example

"Assume that we have the following two log files $\mathbb{L}_1$ and $\mathbb{L}_2$:

Time	Event	Time	Event
1	λ_1	3	λ_3
2	λ_2	6	λ_5
3	λ_3	7	λ_6
4	λ_4		
$\mathbb{L}_1$		$\mathbb{L}_2$	

FIGURE 4.1
Log files $\mathbb{L}_1$ and $\mathbb{L}_2$.

As seen in Figure 4.1, there are two logs one of them has four events and the other has three events. The learning process will start by learning log $\mathbb{L}_1$ and then it follows by learning $\mathbb{L}_2$ before learing $\mathbb{L}_1$, $\Lambda = \emptyset$ and after the learning process $\Lambda = \{\lambda_1, \lambda_2, \lambda_3, \lambda_4\}$. The dependency matrix $\mathbb{D}$ after learning $\mathbb{L}_1$ is as follows:

$$\mathbb{D} = \begin{bmatrix} 0 & 0 & 0 & 0 \\ 1 & 0 & 0 & 0 \\ 0 & 1 & 0 & 0 \\ 0 & 0 & 1 & 0 \end{bmatrix}$$

Dependency matrix $\mathbb{D}$ represents the correlation between events. If event λ_y depends on event λ_x then $\mathbb{D}(y,x) = 1$ otherwise $\mathbb{D}(y,x) = 0$. As seen from log $\mathbb{L}_1$, λ_1 occurs first and therefore, it does not depend on any other event. This is why row $\mathbb{D}(1)$ is all zeros. Event λ_2 however, occurs after event λ_1 and that is why there is a possibility that event λ_2 depends on event λ_1. For the given lof $\mathbb{L}_1$, there is no reason to think that λ_2 does not depend on λ_1. After learning more log files, this assumption might change. Similarly, event λ_3 depends on event λ_2 and event λ_4 depends on event λ_3.

The process continues by learning log file $\mathbb{L}_2$. In $\mathbb{L}_2$, the set of events λ changes = be $\Lambda = \Lambda \cup \lambda_5$. Now updating the dependncy matrix is needed and it goes as follows: for λ_3, it occured this time as the very first event, this means that in log $\mathbb{L}_1$ the event λ_3 occured after λ_1 and λ_2 but it does not depend on any of them. The occurence of λ_3 updates the dependency matrix $\mathbb{D}$ as follows:

$$\mathbb{D} = \begin{bmatrix} 0 & 0 & 0 & 0 & 0 \\ 1 & 0 & 0 & 0 & 0 \\ 0 & 0 & 0 & 0 & 0 \\ 0 & 0 & 1 & 0 & 0 \\ 0 & 0 & 0 & 0 & 0 \end{bmatrix}$$

It is worth noting that the dimensions of the dependency matrix $\mathbb{D}$ has changed to be 5×5 after adding λ_5 to Λ. The row $\mathbb{D}(5)$ is all zeros since the process still did not learn about λ_5.

Now the process learns that Event λ_5 happens after λ_3. So far it is the only fact we know about λ_5 and therefore the assumption would be that λ_5 depends on λ_3. This brings the dependency matrix $\mathbb{D}$ to be:

$$\mathbb{D} = \begin{bmatrix} 0 & 0 & 0 & 0 & 0 \\ 1 & 0 & 0 & 0 & 0 \\ 0 & 0 & 0 & 0 & 0 \\ 0 & 0 & 1 & 0 & 0 \\ 0 & 0 & 1 & 0 & 0 \end{bmatrix}$$

lastly, the process learns from log $\mathbb{L}_2$ about event λ_2. In log $\mathbb{L}_2$, λ_2 occurrs without the occurrence of λ_1 which means that the assumption that was previously made after learning log $\mathbb{L}_1$ no longer holds. Also since λ_2 did not depend on λ_5 or λ_3 in log $\mathbb{L}_1$, this means that λ_2 does not depend on any of them. This brings the dependency matrix to be as follows:

$$\mathbb{D} = \begin{bmatrix} 0 & 0 & 0 & 0 & 0 \\ 0 & 0 & 0 & 0 & 0 \\ 0 & 0 & 0 & 0 & 0 \\ 0 & 0 & 1 & 0 & 0 \\ 0 & 0 & 1 & 0 & 0 \end{bmatrix}$$

From the last update of the dependency matrix, we can conclude that the only depedencies in those logs are λ_4 depending on λ_3 and λ_5 depending on λ_3.

The correlation between events is defined by six square matrices each one for an operator. The dimension of those metrices are $\|\Lambda\| \times \|\Lambda\|$. The **and** relationship matrix, **or** relationship matrix, **XOR** relationship matrix, **implication** relationship matrix, **partial** parallelism matrix, and **full** parallelism matrix are denoted respectively as follows: $\mathbb{R}_\wedge, \mathbb{R}_\vee, \mathbb{R}_\oplus, \mathbb{R}_\rightarrow, \mathbb{R}_{\dashrightarrow} \mathbb{R}_\leftrightarrow$, and $\mathbb{R}_\Leftrightarrow$. The dimesntions of each and every one of those metrices is $\|\Lambda\| \times \|\Lambda\|$ as mentioned earlier. We start with those metrices all initialized with zeros. We also have the following vectors $\mathbb{V}_\mapsto$, $\mathbb{V}_\frown$, and $\mathbb{V}_\circlearrowleft$ The dimention of every one of those vectors is an **upper bound** of the permutation function:

$$\|\mathbb{V}\| = \int_{k=1}^{\|\Lambda\|} \frac{\|\Lambda\|!}{(\|\Lambda\| - k)!} \partial k \tag{4.14}$$

where $x!$ is the factorial of x" (Zayoud et al., 2019a).

• β Approach for Frequent Correlations in Events Log

If actions or events in a log are repeated frequently, or even if those actions also belong to some types of the defined logical relationships, hence a "Frequency Vector" $\mathbb{F}$ will be established from each action in that log that may occur frequently $\mathbb{F}$. A tuple as "(λ, c)" where "c" is a count of action's execution named for example λ, and the values of frequency are traced with the following vectors (Zayoud et al., 2019a):

"Frequency vector"	"Description"
"$\mathbb{F}$	$\forall \lambda \in \Lambda, \|\lambda\|$".
"$\mathbb{F}^\wedge$	$\forall \lambda_i \in \Lambda, \forall \lambda_j \in \Lambda, \|\lambda_i \wedge \lambda_j\|$".
"$\mathbb{F}^\vee$	$\forall \lambda_i \in \Lambda, \forall \lambda_j \in \Lambda, \|\lambda_i \vee \lambda_j\|$".
"$\mathbb{F}^\rightarrow$	$\forall \lambda_i \in \Lambda, \forall \lambda_j \in \Lambda, \|\lambda_i \rightarrow \lambda_j\|$".
"$\mathbb{F}^{\dashrightarrow}$	$\forall \lambda_i \in \Lambda, \forall \lambda_j \in \Lambda, \|\lambda_i \dashrightarrow \lambda_j\|$".
"$\mathbb{F}^\leftrightarrow$	$\forall \lambda_i \in \Lambda, \forall \lambda_j \in \Lambda, \|\lambda_i \leftrightarrow \lambda_j\|$".
"$\mathbb{F}^\Leftrightarrow$	$\forall \lambda_i \in \Lambda, \forall \lambda_j \in \Lambda, \|\lambda_i \Leftrightarrow \lambda_j\|$".
"$\mathbb{F}^\frown$	$\forall \lambda_i \in \Lambda, \forall \lambda_j \in \Lambda, \|\lambda_i \frown \lambda_j\|$".
"$\mathbb{F}^\circlearrowleft$	$\forall \lambda_i \in \Lambda, \forall \lambda_j \in \Lambda, \|\lambda_i \circlearrowleft \lambda_j\|$".

• β As Probabilistic Approach

"After calculating frequencies, probability is straight forward and they are calculated as follows:

$$\mathbb{P}(\lambda) = \frac{\mathbb{F}(\lambda)}{\int_{k=1}^{k=\|\Lambda\|} \mathbb{F}(\lambda_k)} \partial k \qquad (4.15)$$

The probability of two or more events to be joined by relation Θ is as follows:

$$\mathbb{P}(\lambda_i \theta \lambda_j) = \frac{\mathbb{F}^\theta(\lambda_i \theta \lambda_j)}{\int_{k=1}^{k=\|\Lambda\|} \mathbb{F}(\lambda_k)} \partial k \qquad (4.16)$$

where θ could be any operator defined in section 4.6.

Algorithm part 2 of the \$beta Algorithm that is presneted in Appendix A calculates $\mathbb{R}_\theta$, $\mathbb{V}_\theta$, and $\mathbb{F}^\theta$. The input is the log file $\mathbb{L}$, dependency matrix $\mathbb{D}$ and the set of learned events Λ and the output is $\mathbb{R}_\theta$, $\mathbb{V}_\theta$, and $\mathbb{F}^\theta$ wgere θ is any relation described in Section 4.6" (Zayoud et al., 2019a).

4.7 PREDICTION TOWARD KNOWLEDGE

A model of any process in industry is building blocks for predicating and identifying that system toward gaining valuable information that builds a base for the desired knowledge in that system. When some levels of knowledge are achieved many straight rules can be applied to predict events in real life applications.

In the Figure 4.2, the workflow of the proposed algorithm is presented where the data files are first analyzed and then the knowledge discovery starts, with the learning of the events, dependencies, probabilities, and frequencies all will build the workflow model to reach at the end what is called a predict model which describes the current process clearly.

The below Symbols are predictions rules for gaining knowledge:
- $\text{Occurs}(\lambda_x)$
- $\text{And}(\lambda_x, \lambda_y)$
- $\text{Or}(\lambda_x, \lambda_y)$
- $\text{Xor}(\lambda_x, \lambda_y)$
- $\text{Depends}(\lambda_x, \lambda_y)$
- $\text{Follows}(\lambda_x, \lambda_y)$
- $\text{PartialParallel}(\lambda_x, \lambda_y)$
- $\text{FullParallel}(\lambda_x, \lambda_y)$
- $\text{Sequence}(\lambda_x, \ldots, \lambda_y)$
- $\text{RepeatedSequence}(\lambda_x, \ldots, \lambda_y)$
- $\text{Loop}(\lambda_x, \ldots, \lambda_y)$

The constraints of the knowledgebase are desribed as the following:

$$"\forall x, \forall y, \text{And}(\lambda_x, \lambda_y)" \wedge \text{Occurs}(\lambda_x) \rightarrow \text{Occurs}(\lambda_y) \qquad (4.17)$$

$$\forall x, \forall y, \text{Or}(\lambda_x, \lambda_y) \rightarrow \text{Occurs}(\lambda_y) \vee \text{Occurs}(\lambda_x) \qquad (4.18)$$

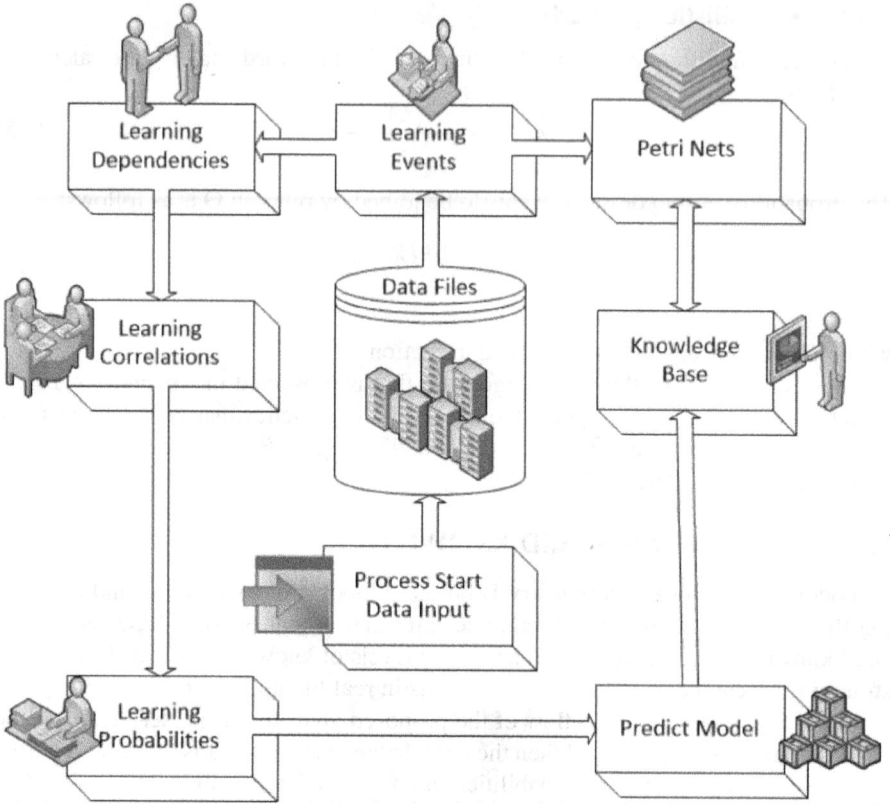

FIGURE 4.2
The β algorithm framework flowchart.

$$\forall x, \forall y, \text{Xor}(\lambda_x, \lambda_y) \rightarrow (\text{Occurs}(\lambda_y) \wedge\, ! \,\text{Occurs}(\lambda_x)) \vee (\text{Occurs}(\lambda_y) \wedge \text{Occurs}(\lambda_x)) \tag{4.19}$$

$$\forall x, \forall y, \text{Depends}(\lambda_x, \lambda_y) \leftrightarrow \text{Occurs}(\lambda_x) \rightarrow \text{Occurs}(\lambda_y) \tag{4.20}$$

$$\forall x, \forall y, \exists t_y, \exists t_x, \text{Follows}(\lambda_x, \lambda_y) \rightarrow \text{Occurs}(\lambda_x) \wedge \text{Occurs}(\lambda_y) \wedge t_y \geq t_x \tag{4.21}$$

$$\forall x, \forall y, \exists t_y, \exists t_x, \text{PartialParallel}(\lambda_x, \lambda_y) \rightarrow \text{Occurs}(\lambda_x) \tag{4.22}$$

$$\wedge \text{Occurs}(\lambda_y) \wedge (t_y \leq t_x \leq t_y + \tau_y) \vee (t_x \leq t_y \leq t_x + \tau_x) \tag{4.23}$$

$$\forall x, \forall y, \exists t_y, \exists t_x, \text{FullParallel}(\lambda_x, \lambda_y) \rightarrow \text{Occurs}(\lambda_x) \wedge \text{Occurs}(\lambda_y) \wedge (t_y = t_x) \tag{4.24}$$

$$\forall x, \forall y, \forall z, \text{Sequence}(\lambda_x, \lambda_y, \lambda_z) \rightarrow \text{Depends}(\lambda_y, \lambda_x) \wedge \text{Depends}(\lambda_z, \lambda_y) \tag{4.25}$$

$$\forall \, \text{Sequence}(\lambda_x, \ldots, \lambda_z), \forall W, \text{Sequence}(\lambda_x, \ldots, \lambda_z, \lambda_w) \rightarrow \text{Depends}(\lambda_w, \lambda_z) \tag{4.26}$$

$$\forall \text{RepeatedSequence}(\lambda_x, \ldots, \lambda_z) \rightarrow \text{Sequence}(\lambda_x, \ldots, \lambda_z) \wedge$$

$$\| \, \text{Sequence}(\lambda_x, \ldots, \lambda_z \in \mathbb{L} \| \geq 2 \tag{4.27}$$

$$\wedge \exists \lambda_k \notin \text{RepeatedSequence}(\lambda_x, \ldots, \lambda_z), t_k \geq t_r$$

and "t_r is a time instant between the occurrences of two instances of the sequence. $\mathbb{L}$ is the log file under study" (Zayoud et al., 2019a).

$$\forall \text{Loop}(\lambda_x, \ldots, \lambda_z) \rightarrow \text{Sequence}(\lambda_x, \ldots, \lambda_z) \tag{4.28}$$

$$\wedge \| \text{Sequence}(\lambda_x, \ldots, \lambda_z \in \mathbb{L}\| \geq 2 \wedge \not\exists \lambda_k \notin \text{RepeatedSequence}(\lambda_x, \ldots, \lambda_z), t_k \geq t_r \tag{4.29}$$

Equations from 4.17 to 4.29 describe how we can infer and predict the coming events based on the currently occurring events. After learning the dependency matrix $\mathbb{D}$, the Relationship matrices $\mathbb{R}_\theta$ and vectors $\mathbb{V}_\theta$, the system is ready to feed knowledge base learned **facts** to predict what is coming in the near future. We also learn the frequencies $\mathbb{F}_\theta$ and the probabilities of those predicates to be true, this way we can make an accurate prediction to what is coming. This is illustrated in a future example for healthcare (Zayoud et al., 2019a).

4.8 APPLYING THE β ALGORITHM IN HEALTHCARE

In this section, healthcare log files are studied and set of main events are concluded with their corresponding resources, times, and frequencies. However, Learning the events, their dependencies, correlations and their probabilities, is the building block of approaching a process model. As seen early in this chapter, in literature and survey sections, the process mining methods have many difficulties in approaching an efficient process model. Hence, the proposed framework is to overcome these limitations and provides an efficient knowledge base that predicts a coherent process model as proved in Zayoud et al. (2019a).

To extract logs, many data files needed where the records of unclassified data stored that is in the available (IS) or various database engines, is the key to extract those log files. In reference to healthcare systems, hospitals are important part in this systems, and the log files are extracted from multiple records that contain data of those hospitals. As an example of extracted log file of hospital data, the following contents of log file are discovered according to various records in a hospital and based on the paper (Oueida et al., 2017), activities can be classified as:

- "Arrival: arrival of patient to the hospital".
- "Data Collection: Process of collecting patient information".
- "Check Urgent: process of defining the severity level of an arriving patient".
- "Decide: Process of deciding whether the patient case is urgent or not".
- "Examination: process of examination and checkup".
- "Waiting room: Non-urgent patients are referred to a waiting room".
- "Treatment: process of pre-checkup by a nurse until a doctor is available".
- "Release resource: once the checkup is over, the resource in charge should be released".
- "Examination: process performed by a doctor in order to examine the arriving patient".
- "Billing Process".

- "Extra Facilities: Decision whether the patient needs extra facilities or no."
- "Radiology: represents the radiology unit where patients do X-rays, Citi-Scan, MRI, or other types of imaging as requested by doctors".
- "Treatment: During this process, patient undergoes the required treatment as prescribed by the doctor".

All the mentioned and quoted events in addition to the resources are studied and publish in my paper (Zayoud et al., 2019a). Where resources are considered as:

▶ "Doctor".
▶ "Nurse".
▶ "Transporter".
▶ "Radiology technician".
▶ "Accountant".
▶ "Non-Human resources like: Medical equipment's, beds, chairs etc." (Zayoud et al., 2019a).

The mentioned resources and activities are in Figure 4.3.

Data files of patients in addition to the new approach of this book both provide theoretically what are names logs of events in those hospitals, an overview of a sample log of events is prepared and described as the following:

TABLE 4.3
Log of Events 1

Time	Event	Duration
7:45	Arrival	5
7:53	Checking Urgent	15
8:10	Waiting room	35
8:59	Doctor Decide	25
9:30	Examination	60
10:45	Treatment	50
11:40	Release resource	5
12:00	Billing	20

Log 2 is described below and it is the second sample of log files but for different patient than the one in log 1 with same period of time.

Last example is handling different patients' actions for a week, knowing that information of time and duration is expressed by minutes.

The coming dependencies metric is given in log 1 to log 3 respectively, and its algorithm is expressed as a new solution. The "dependency matrix" for the 8 events

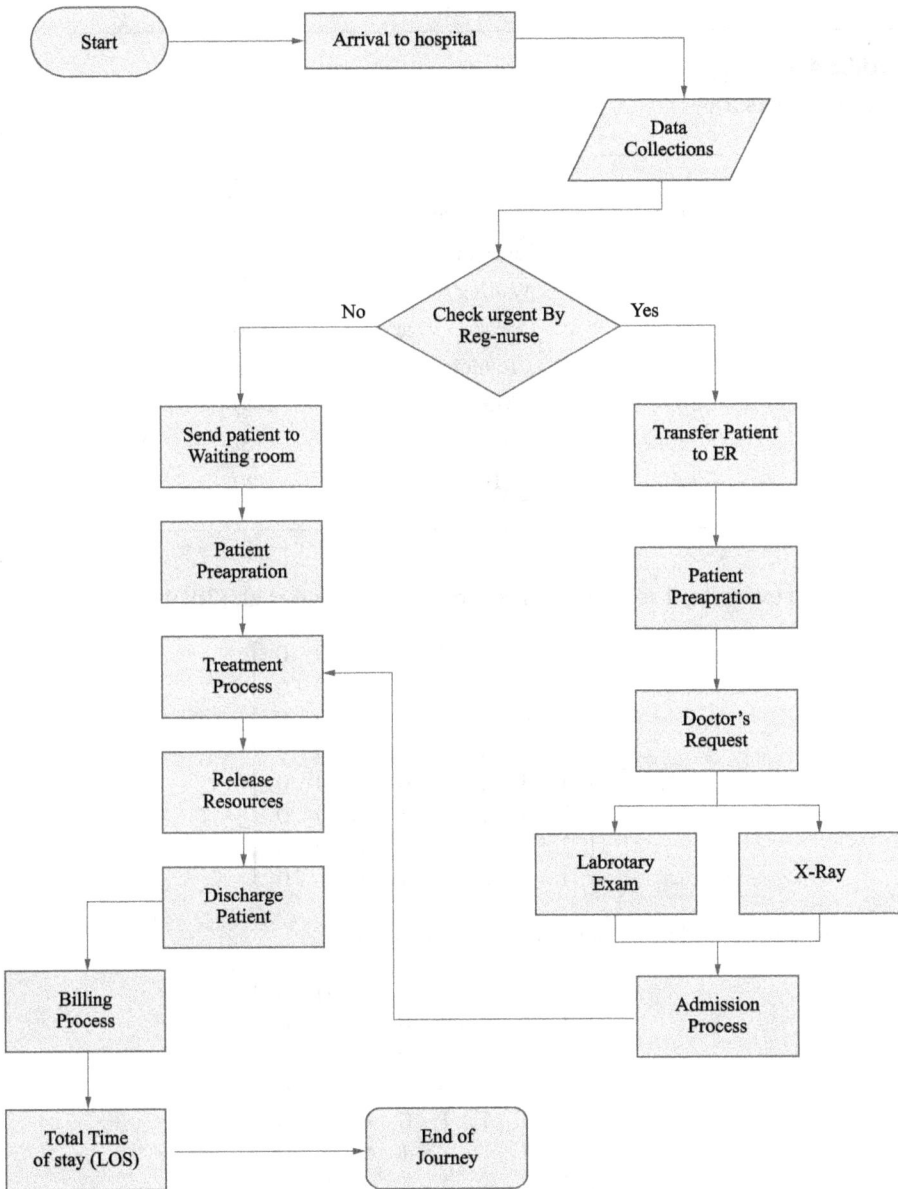

FIGURE 4.3
Process flow in particular hospital.

TABLE 4.4

Log of Events 2

Time	Event	Duration
9:05	Arrival	10
9:06	Data Collection	5
9:15	Waiting room	70
9:59	Doctor Decide	20
9:30	Radiology	40
10:45	Treatment	40
11:40	Release resource	5
12:00	Billing	20

presents in event log 1 is as follows according to Zayoud et al. (2019a):

$$
\mathbb{D} = \begin{bmatrix}
0 & 0 & 0 & 0 & 0 & 0 & 0 & 0 \\
1 & 0 & 0 & 0 & 0 & 0 & 0 & 0 \\
1 & 1 & 0 & 0 & 0 & 0 & 0 & 0 \\
1 & 1 & 1 & 0 & 0 & 0 & 0 & 0 \\
1 & 1 & 1 & 1 & 0 & 0 & 0 & 0 \\
1 & 1 & 1 & 1 & 1 & 0 & 0 & 0 \\
1 & 1 & 1 & 1 & 1 & 1 & 0 & 0 \\
1 & 1 & 0 & 1 & 1 & 1 & 1 & 0
\end{bmatrix}
$$

The "Dependency Matrix" in second log is as follows:

$$
\mathbb{D} = \begin{bmatrix}
0 & 0 & 0 & 0 & 0 & 0 & 0 & 0 \\
1 & 0 & 0 & 0 & 0 & 0 & 0 & 0 \\
1 & 1 & 0 & 0 & 0 & 0 & 0 & 0 \\
1 & 1 & 1 & 0 & 0 & 0 & 0 & 0 \\
1 & 1 & 1 & 1 & 0 & 0 & 0 & 0 \\
1 & 1 & 1 & 1 & 1 & 0 & 0 & 0 \\
1 & 1 & 1 & 1 & 1 & 1 & 0 & 0 \\
1 & 1 & 0 & 1 & 1 & 1 & 1 & 0
\end{bmatrix}
$$

TABLE 4.5
Log of Events 3

Time	Event	Duration	Patient ID
10:15	Arrival	5	12496
10:20	Data Collection	10	12496
10:25	Waiting room	30	12496
10:25	Arrival	10	232004
10:25	Arrival	5	451233
10:30	Data Collection	10	232004
10:35	Waiting room	20	232004
10:35	Data Collection	5	451233
10:45	Waiting room	35	451233
10:59	Doctor Decide	15	12496
11:30	Radiology	40	12496
11:15	Doctor Decide	15	232004
11:30	Examination	60	232004
12:35	Treatment	130	12496
11:20	Doctor Decide	25	451233
11:35	Treatment	10	451233
01:35	Hospital Admission	45	232004
11:45	Release resource	10	451233
12:00	Billing	20	451233
02:55	Release resource	10	12496
03:00	Billing	20	12496
02:15	Treatment	3500	232004
12:00	Release resource	60	232004
01:05	Billing	35	232004

The "Dependency Matrix" for the the third log of nine activities as follows:

$$
\mathbb{D} = \begin{bmatrix}
0 & 0 & 0 & 0 & 0 & 0 & 0 & 0 & 0 \\
1 & 0 & 0 & 0 & 0 & 0 & 0 & 0 & 0 \\
1 & 1 & 0 & 0 & 0 & 0 & 0 & 0 & 0 \\
1 & 1 & 1 & 0 & 0 & 0 & 0 & 0 & 0 \\
1 & 1 & 1 & 1 & 0 & 0 & 0 & 0 & 0 \\
1 & 1 & 1 & 1 & 0 & 0 & 0 & 0 & 0 \\
1 & 1 & 1 & 1 & 1 & 1 & 0 & 0 & 0 \\
1 & 1 & 1 & 1 & 1 & 1 & 1 & 0 & 0 \\
1 & 1 & 0 & 1 & 1 & 1 & 1 & 1 & 0
\end{bmatrix}
$$

The frequency of the events mentioned in log 1 can be seen as the following:

Frequency Vector	Description
$\mathbb{F}^\wedge$	Arrival and Check urgent.
$\mathbb{F}^\vee$	Waiting Room, Doctoct Decide.
$\mathbb{F}^\oplus$	Doctor Decide, Billing.
$\mathbb{F}^\rightarrow$	Check urgent depends on Arrival.
$\mathbb{F}^{--\rightarrow}$	Waiting follows Check urgent.
$\mathbb{F}^\leftrightarrow$	Arrival, Waiting.
$\mathbb{F}^\Leftrightarrow$	None of log1 events.
$\mathbb{F}^\frown$	Arrival, Data, Waiting.
$\mathbb{F}^\circlearrowleft$	All events in log 1.

• Frequency of Events and Their Co-relation

The frequency of occurrence of events and the co-relations with events are kept in a frequency vector $\mathbb{F}$. Every element of $\mathbb{F}$ is an event or a set of events and their occurrence. That tupple is in the form of (λ, c) where c is the number of occurrences of event λ. An example of $\mathbb{F}$ would be as follows after learning Log event 1 over one month as in Zayoud et al. (2019a):

$$\mathbb{F} = \{(\text{“Arrival”}, \text{“2”}), (\text{“Checking urgent”}, \text{“1”}), (\text{“Data Collection”}, \text{“1”}),$$
$$(\text{“DoctorDecide”}, \text{“2”}), (\text{“Radiology”}, \text{“1”}), (\text{“Treatment”}, \text{“2”}),$$
$$(\text{“Release Resource”}, \text{“8”}), (\text{“Billing”}, \text{“2”})\}$$
$$\mathbb{F}^\wedge = \{(\text{“Arrival”} \wedge \text{“CheckingUrgent”}, \text{“1”})\}$$
$$\mathbb{F}^\vee = \{\bigcup_{k=1}^{\|\Lambda\|} P_k(\Lambda, \Lambda)\} \text{ where } k \text{ is permutation length.}$$
$$\mathbb{F}^\oplus = \{(\text{“Arrival”} \oplus \text{“Billing”}, \text{“2”}), (\text{“Radiology”} \oplus \text{“Treatment”}, \text{“1”}),$$
$$(\text{“Checkurgent”} \oplus \text{“Waiting”}, 1),\text{”}$$
$$\mathbb{F}^\rightarrow = \{(\text{Check Urgent} \rightarrow \text{Arrival}, \text{“1”}),$$
$$(\text{“Treatment”} \rightarrow \text{“Radiology”}, \text{“1”}),$$
$$(\text{“Billing”} \rightarrow \text{“Treatment”}, \text{“2”}),$$
$$(\text{“Doctor Decide”} \rightarrow \text{“Check Urgent”}, \text{“1”}),$$
$$(\text{“Data Collection”} \rightarrow \text{“Arrival”}, \text{“1”})\}$$
$$\mathbb{F}^{--\rightarrow} = \{(\text{“Waiting”} --\rightarrow \text{“Check Urgent”}, \text{“1”}),$$
$$(\text{“Treatment”} --\rightarrow \text{“Doctor Decide”}, \text{“2”}),$$
$$(\text{“Radiology”} --\rightarrow \text{“Doctor Decide”}, \text{“1”}),$$
$$(\text{“Release Resource”} --\rightarrow \text{“Treatment”}, \text{“2”})\}$$
$$\mathbb{F}^\leftrightarrow = \{(\text{“Arrival”} \leftrightarrow \text{“Check Urgent”}, \text{“1”})\}$$
$$\mathbb{F}^\Leftrightarrow = \{\emptyset\}$$
$$\mathbb{F}^\mapsto = \{(\text{“Arrival”} \mapsto \text{“Data Collection”}, \text{“1”}),$$
$$(\text{“DoctorDecide”} \mapsto \text{“Radiology”}, \text{“1”}),$$
$$(\text{“DoctorDecide”} \mapsto \text{“Treatment”} \mapsto$$

Release Resource, "2")}
$$\mathbb{F}^\frown = \{(\text{"Arrival"} \mapsto \text{"Data Collection"} \mapsto \text{Waiting}, 1)\}$$
$$\mathbb{F}^\circlearrowleft = \{(\text{"Arrival"} \circlearrowleft \text{"DataCollection"}$$
$$\circlearrowleft \text{"DoctorDecide"} \circlearrowleft \text{"Treatment"}, \text{"1"})\}$$

The actions with frequent occurrence in the third Log are for more than one patient are described as the following:

"Frequency vector"	"Description"
"$\mathbb{F}$"$^\wedge$	"Arrival" "Treatment".
"$\mathbb{F}$"$^\vee$	"Radiology", "Examniation".
"$\mathbb{F}$"$^\oplus$	"Doctor Decide", "Billing".
"$\mathbb{F}$"$^\rightarrow$	"Data collection" depends on "Arrival".
"$\mathbb{F}$"$^{--\rightarrow}$	"Waiting" follows "Data Collection".
"$\mathbb{F}$"$^\leftrightarrow$	"Arrival", "Waiting".
"$\mathbb{F}$"$^\Leftrightarrow$	"None of log1 events".
"$\mathbb{F}$"$^\frown$	"Arrival", "Data collection", "Waiting".
"$\mathbb{F}$"$^\circlearrowleft$	"Actions of 1st patient and 3rd patient only.

The table of "Frequency" relations of third log of events is implemented as an example of "$\mathbb{F}$" which is changed after learning the third log for various patients:

$$\text{"}\mathbb{F}\text{"} = \{(\text{"Arrival"}, 3), (\text{"Checking urgent"}, 1), (\text{"Data Collection"}, 2),$$
$$(\text{"DoctorDecide"}, 3), (\text{"Radiology"}, 1), (\text{"Examination"}, 1),$$
$$(\text{"Release Resource"}, 12), (\text{"Billing"}, 3)\}$$
$$\mathbb{F}^\wedge = \{(\text{"Arrival"} \wedge \text{"CheckingUrgent"}, 1)\}$$
$$\mathbb{F}^\vee = \{\bigcup_{k=1}^{\|\Lambda\|} P_k(\Lambda, \Lambda)\} \text{ where } k \text{ is permutation length.}$$
$$\mathbb{F}^\oplus = \{(\text{"Arrival"} \oplus \text{"Billing"}, 3), (\text{"Radiology"} \oplus \text{"Treatment"}, 1),$$
$$(\text{"Checkurgent"} \oplus \text{"Waiting"}, 1),\}$$
$$\mathbb{F}^\rightarrow = \{(\text{"Check Urgent"} \rightarrow \text{"Arrival"}, 1),$$
$$(\text{"Treatment"} \rightarrow \text{"Radiology"}, 1),$$
$$(\text{"Billing"} \rightarrow \text{"Treatment"}, 3),$$
$$(\text{"Doctor Decide"} \rightarrow \text{"Check Urgent"}, 1),$$
$$(\text{"Data Collection"} \rightarrow \text{"Arrival"}, 2)\}$$
$$\mathbb{F}^{--\rightarrow} = \{(\text{Waiting} \dashrightarrow \text{"Check Urgent"}, 1),$$
$$(\text{"Treatment"} \dashrightarrow \text{"Doctor Decide"}, 3),$$
$$(\text{"Radiology"} \dashrightarrow \text{"Doctor Decide"}, 1),$$
$$(\text{"Release Resource"} \dashrightarrow \text{"Treatment"}, 3)\}$$
$$\mathbb{F}^\leftrightarrow = \{(\text{"Arrival"} \leftrightarrow \text{"Check Urgent"}, 1)\}$$
$$\mathbb{F}^\Leftrightarrow = \{(\text{"Arrival patient"}2 \Leftrightarrow \text{"Arrival Patient"}3\}$$
$$\mathbb{F}^\mapsto = \{(\text{"Arrival"} \mapsto \text{"Data Collection"}, 2),$$
$$(\text{"DoctorDecide"} \mapsto \text{"Radiology"}, 1),$$
$$(\text{"DoctorDecide"} \mapsto \text{"Treatment"} \mapsto$$
$$\text{"Release Resource"}, 3)\}$$

$$\mathbb{F}^\frown = \{(\text{"Arrival"} \mapsto \text{"Data Collection"} \mapsto \text{"Waiting"}, 3)\}$$
$$\mathbb{F}^\circlearrowleft = \{(\text{"Arrival"} \circlearrowleft \text{"DataCollection"}$$
$$\circlearrowleft \text{"DoctorDecide"} \circlearrowleft \text{"Treatment"}, 3)\}$$

The co-relations of the eight activities exist in the first log file with a condition that all activities are ordered respectively as seen in the first log in the figure, knowing that the logic value "0" means there is no co-relation between two events and the logic value of "1" means a co-relation is existed between the two events in that cell of the matrix.

From Equation 4.3, $\mathbb{R}_\wedge$ is as follows:

$$\mathbb{R}_\wedge = \begin{bmatrix} 1 & 1 & 0 & 0 & 0 & 0 & 0 & 0 \\ 1 & 1 & 1 & 1 & 0 & 0 & 0 & 0 \\ 0 & 1 & 1 & 1 & 0 & 0 & 0 & 0 \\ 0 & 0 & 0 & 1 & 1 & 1 & 0 & 0 \\ 0 & 0 & 0 & 1 & 1 & 1 & 0 & 0 \\ 0 & 1 & 0 & 1 & 1 & 1 & 1 & 0 \\ 0 & 1 & 0 & 0 & 1 & 1 & 1 & 1 \\ 0 & 0 & 0 & 0 & 0 & 0 & 1 & 1 \end{bmatrix}$$

Equation 4.3 gives a super set of the "AND equation".

"Equation 4.4", presents $\mathbb{R}_\vee$ as follows:

$$\mathbb{R}_\vee = \begin{bmatrix} 1 & 1 & 1 & 1 & 0 & 1 & 1 & 0 \\ 1 & 1 & 1 & 1 & 1 & 1 & 1 & 0 \\ 1 & 1 & 1 & 1 & 0 & 0 & 0 & 0 \\ 1 & 1 & 1 & 1 & 1 & 1 & 1 & 0 \\ 1 & 1 & 0 & 1 & 1 & 1 & 1 & 0 \\ 1 & 1 & 0 & 1 & 1 & 1 & 1 & 1 \\ 1 & 1 & 0 & 1 & 1 & 1 & 1 & 1 \\ 0 & 0 & 0 & 0 & 0 & 1 & 1 & 1 \end{bmatrix}$$

"Equation 4.4" provides "OR equation" with a super set.

While "Equation 4.6", $\mathbb{R}_\oplus$ is shown as:

$$\mathbb{R}_\oplus = \begin{bmatrix} 0 & 0 & 0 & 0 & 0 & 0 & 1 & 1 \\ 0 & 0 & 0 & 0 & 1 & 0 & 1 & 1 \\ 0 & 0 & 0 & 0 & 0 & 1 & 1 & 1 \\ 0 & 0 & 0 & 0 & 0 & 0 & 0 & 1 \\ 1 & 1 & 0 & 0 & 0 & 0 & 1 & 1 \\ 0 & 0 & 0 & 0 & 0 & 0 & 0 & 1 \\ 0 & 0 & 0 & 0 & 0 & 1 & 0 & 0 \\ 1 & 1 & 1 & 1 & 1 & 1 & 1 & 0 \end{bmatrix}$$

Also "Equation 4.6" applys on the "XOR equation" a super set.

"Equation 4.7" of "$\mathbb{R}_\rightarrow$" is presented as:

$$
\mathbb{R}_\rightarrow = \begin{bmatrix}
1 & 0 & 0 & 0 & 0 & 0 & 0 & 0 \\
1 & 1 & 0 & 0 & 0 & 0 & 0 & 0 \\
1 & 1 & 1 & 1 & 0 & 0 & 0 & 0 \\
1 & 1 & 1 & 1 & 1 & 0 & 0 & 0 \\
1 & 1 & 1 & 1 & 1 & 0 & 1 & 0 \\
1 & 1 & 0 & 1 & 1 & 1 & 1 & 0 \\
0 & 1 & 0 & 1 & 0 & 1 & 1 & 0 \\
1 & 0 & 0 & 1 & 1 & 1 & 1 & 1
\end{bmatrix}
$$

Also "Equation 4.8" supplies a super set,
from "Equation 4.8, $\mathbb{R}_{--\rightarrow}$," that is described as follows:

$$
\mathbb{R}_{--\rightarrow} = \begin{bmatrix}
0 & 0 & 0 & 0 & 0 & 0 & 0 & 0 \\
1 & 0 & 0 & 0 & 0 & 0 & 0 & 0 \\
1 & 1 & 0 & 0 & 0 & 0 & 0 & 0 \\
0 & 1 & 0 & 0 & 1 & 0 & 0 & 0 \\
0 & 1 & 0 & 1 & 0 & 0 & 0 & 0 \\
0 & 1 & 0 & 1 & 1 & 0 & 0 & 0 \\
0 & 0 & 0 & 0 & 0 & 1 & 0 & 0 \\
0 & 0 & 0 & 0 & 0 & 0 & 1 & 0
\end{bmatrix}
$$

In "Equation 4.8" a super set is provided by following relation equation.
Here "Equation 4.9, $\mathbb{R}_{\leftrightarrow}$" is expressed as:

$$
\mathbb{R}_{\leftrightarrow} = \begin{bmatrix}
1 & 1 & 1 & 0 & 0 & 0 & 0 & 0 \\
1 & 1 & 1 & 1 & 1 & 1 & 0 & 0 \\
1 & 1 & 1 & 1 & 0 & 0 & 0 & 0 \\
0 & 1 & 1 & 1 & 1 & 1 & 0 & 0 \\
0 & 0 & 0 & 1 & 1 & 0 & 0 & 0 \\
0 & 1 & 1 & 1 & 1 & 1 & 0 & 0 \\
0 & 0 & 0 & 0 & 0 & 1 & 1 & 0 \\
0 & 0 & 0 & 0 & 0 & 0 & 0 & 1
\end{bmatrix}
$$

"Equation 4.9 also provides a super set of the "Partial Parallelism" equation.
"Equation 4.10, $\mathbb{R}_{\leftrightarrow}$" is expressed as:

$$
\mathbb{R}_{\leftrightarrow} = \begin{bmatrix}
0 & 0 & 0 & 0 & 0 & 0 & 0 & 0 \\
0 & 0 & 0 & 0 & 0 & 0 & 0 & 0 \\
0 & 0 & 0 & 0 & 0 & 0 & 0 & 0 \\
0 & 0 & 0 & 0 & 0 & 0 & 0 & 0 \\
0 & 0 & 0 & 0 & 0 & 0 & 0 & 0 \\
0 & 0 & 0 & 0 & 0 & 0 & 0 & 0 \\
0 & 0 & 0 & 0 & 0 & 0 & 0 & 0 \\
0 & 0 & 0 & 0 & 0 & 0 & 0 & 0
\end{bmatrix}
$$

"Equation 4.10" is also a super set of the "Full Parallelism" equation.

"Equation 4.11, $\mathbb{R}_\mapsto$" is given as the following:

$$\mathbb{R}_\mapsto = \begin{bmatrix} 1 & 1 & 0 & 0 & 0 & 0 & 0 & 0 \\ 1 & 1 & 0 & 0 & 0 & 0 & 0 & 0 \\ 1 & 0 & 1 & 0 & 0 & 0 & 0 & 0 \\ 1 & 1 & 0 & 1 & 0 & 0 & 0 & 0 \\ 0 & 0 & 0 & 0 & 1 & 0 & 0 & 0 \\ 0 & 0 & 0 & 0 & 0 & 1 & 0 & 0 \\ 0 & 0 & 0 & 0 & 0 & 0 & 0 & 0 \\ 0 & 0 & 0 & 0 & 0 & 0 & 0 & 0 \end{bmatrix}$$

A super set is provided by "Equation 4.11" from the "Sequence" equation.

"Equation 4.12, $\mathbb{R}_\frown$" is also following the below expression:

$$\mathbb{R}_\frown = \begin{bmatrix} 1 & 1 & 0 & 0 & 0 & 0 & 0 & 0 \\ 1 & 1 & 0 & 0 & 0 & 0 & 0 & 0 \\ 1 & 0 & 1 & 0 & 0 & 0 & 0 & 0 \\ 1 & 1 & 0 & 1 & 0 & 0 & 0 & 0 \\ 0 & 0 & 0 & 0 & 1 & 0 & 0 & 0 \\ 0 & 0 & 0 & 0 & 0 & 1 & 0 & 0 \\ 0 & 0 & 0 & 0 & 0 & 0 & 0 & 0 \\ 0 & 0 & 0 & 0 & 0 & 0 & 0 & 0 \end{bmatrix}$$

Another super set of "Equation 4.12" gis from the "Repeated Sequence" equation.

"Equation 4.13, $\mathbb{R}_\circlearrowleft$" is as follows:

$$\mathbb{R}_\circlearrowleft = \begin{bmatrix} 0 & 0 & 0 & 0 & 0 & 0 & 0 & 0 \\ 0 & 0 & 0 & 0 & 0 & 0 & 0 & 0 \\ 1 & 1 & 1 & 0 & 0 & 0 & 0 & 0 \\ 1 & 1 & 1 & 1 & 0 & 0 & 0 & 0 \\ 0 & 0 & 0 & 0 & 0 & 0 & 0 & 0 \\ 0 & 0 & 0 & 0 & 0 & 0 & 0 & 0 \\ 0 & 0 & 0 & 0 & 0 & 0 & 0 & 0 \\ 1 & 1 & 1 & 1 & 0 & 1 & 1 & 1 \end{bmatrix}$$

Finally, "Equation 4.13" is again a provider for a super set of a "Loop" equation.

The correlations for both logs 1 and 2 are similar, such that "R" = "1" for two activities if there is a match to one of the mentioned rules. The correlations of the events exists, i.e., to have a logic value of "1" for part of activities in the third log and can be seen as the coming example table:

4.9   LEARNING EVENTS PROBABILITIES

"Learning dependency of events means that learning about the events which are related and events that are not related to each other in an event log, this leads to learning probabilities of these events and calculates the likelyhood of any event to happen

TABLE 4.6
Correlations of the Third Log

Correlation		Description	
$\mathbb{R}^{\wedge}$	Arrival	all other events	= 1
$\mathbb{R}^{\vee}$	Doctor	Radiology	= 1
$\mathbb{R}^{\oplus}$	No Arrival	Billing	= 1
$\mathbb{R}^{\rightarrow}$	Data collection	Arrival	= 1
$\mathbb{R}^{--\rightarrow}$	Waiting follows	Data Collection	= 1
$\mathbb{R}^{\leftrightarrow}$	Arrival	Waiting	= 1
$\mathbb{R}^{\Leftrightarrow}$	= 0	for all events	
$\mathbb{R}^{\frown}$	Arrival	Data collection	= 1
$\mathbb{R}^{\circlearrowleft}$	Patient 1	Patient 3	= 1

based on previous knowledge of these events and their dependency, correlations, and frequency calculations. However, the events are defined as independent when two or more events have no effect on the occurance of each other. On other hand, the dependent events are the events that their occurance of one event has an effect on the other. As an example, in our hospital application, the collection of patient data dependes on patient arrival to the hospital.

The following is to study the probability of events for all the relations that are mentioned before in this chapter which are the probability of

- "P ($\wedge$) relation"
- "P ($\vee$) relation"
- "P ($\oplus$) relation"
- "P ($\rightarrow$) relation"
- "P ($--\rightarrow$) relation"
- "P ($\leftrightarrow$) relation"
- "P ($\Leftrightarrow$) relation"
- "P ($\frown$) relation"
- "P ($\circlearrowleft$) relation"

Based on the probability theories equations (Durrett, 2010a), therefore,

For one activity "λ_j" to **"AND"** with an action, the probability of these this case when they have AND relation is

$$P(\lambda_i \wedge \lambda_j) = \frac{\text{Occurance Of}(\lambda_i \wedge \lambda_j)}{\text{Total Number Of Events}} \quad (4.30)$$

For one activity "λ_j" to **"OR"** with an action, the probability of this case when they have "OR" relation is

$$P(\lambda_i \vee \lambda_j) = \text{``}P(\lambda_i) + P(\lambda_j) - P((\lambda_i \cap \lambda_j)\text{''} \quad (4.31)$$

In case of an activity λ_j to **"XOR"** with anotheractivity, hence the calculations of probability in the case of having XOR relation is

$$P(\lambda_i \oplus \lambda_j) = \text{``}1 - P(\lambda_i \cap \lambda_j)\text{''} \quad (4.32)$$

In the situation of action λ_j to **"Depends"** on another action, this situation has a probability actions that are under "dependency relation" as:

$$P(\lambda_i \rightarrow \lambda_j) = \frac{\text{``}P(\lambda_i \cap \lambda_j)\text{''}}{\text{``}P(\lambda_j\text{'')}} \quad (4.33)$$

If activity λ_j to **"Follow"** by anotheractivity, hence "Follow" relationship between activities in a log has a probability as:

$$P(\lambda_i \dashrightarrow \text{``}\lambda_j\text{''}) = \text{``}P(\lambda_i \cap \lambda_j)\text{''} \quad (4.34)$$

If an activity "λ_j" is to **"Partial Parallel"** with another activity, this type of relationship between actvities means a probability expression as following:

$$P(\lambda_i \dashrightarrow \lambda_j) = \text{``}P(\lambda_i \cap \lambda_j)\text{''} \quad (4.35)$$

An activity λ_j is to **"Full Parallel"** to different activity, then the probability of both activities with "Full parallel" relationship is expressed as:

$$P(\lambda_i \leftrightarrow \lambda_j) = \text{``}P(\lambda_i \cap \lambda_j)\text{''} \quad (4.36)$$

The term activities or events such as $\lambda_i, \lambda_j, \ldots, \lambda_n$ that are consider as **Sequence**, the probability of them when they have "Sequence relation" is:

$$P(\lambda_i \leftrightarrow (\lambda_j, \lambda_k) = \frac{P(\lambda_i \cap (\lambda_j \cap \lambda_k))}{P(\lambda_j \cap \lambda_k)} \quad (4.37)$$

For the events $\lambda_i, \lambda_j, \ldots, \lambda_n$ that are consider as **Repeated Sequence**, the probability of these events when they considers as repeated sequence relation is

$$P \curvearrowright (\lambda_i, \lambda_j, \lambda_k) = \frac{\text{Frequency of this sequence}}{\text{Total number of events}} \quad (4.38)$$

For the events $\lambda_i, \lambda_j, \ldots, \lambda_n$ that are consider as **Loop**, the probability of these events when they considers as loop relation is

$$P \circlearrowright (\lambda_i, \lambda_j, \lambda_k) = 0 \quad \text{if} \quad P(\lambda_i) = 0 \quad (4.39)$$

$$P \circlearrowright (\lambda_i, \lambda_j, \lambda_k) = 1 \quad \text{if} \quad P(\lambda_i) = 1 \quad (4.40)$$

4.10 PETRI NETS OF THE β ALGORITHM

Petri Net is one of several mathematical modelling languages that is known as a place/transition (PT) net, as a description of distributed systems. It is a class of discrete event dynamic system. A Petri Net is a directed bipartite graph, in which the nodes represent transitions that are events that may occur, transitions are represented by bars (Petri and Reisig, 2008). Places in Petri Nets are conditions, represented by circles. The directed arcs describe which places are pre- and/or post-conditions for which transitions (signified by arrows). Petri Net is a graphical notation for processes that have choice, iteration, and concurrent execution. However, it is an accurate mathematical definition of their execution semantics based on mathematical theory for process analysis that is will developed. Petri Nets have been studied in depth from many points of view: from their clear semantic to a certain number of possible extensions (such as time, color) (Zayoud et al., 2019a).

Formal definitions of Petri Net are presented, as in Murata (1989a), as the following: Definition 1 (Petri Net). A Petri Net is a tuple (P, T, F) where: P is a finite set of places; T is a finite set of transitions, such that $P \cap T = \phi$, and F (P×T) ∪ (T × P) is a set of directed arcs, called flow relation (Zayoud et al., 2019a).

The "dynamic semantic of a Petri Net" is dependent on the "firing rule": hence what is called as a "transition" fires all its "input places" where places with edges entering into the transition consist of one token with the minimum scale. One token for all its "output places" is established when there is a transition that is fired. However, "Marking" is the distributed tokens over places in an existed net with a specific time. Hence, various behaviors and models can be extracted With this semantic. The "Sequence template" expresses the "Causal Dependency" among two events, while the "AND template" describes the simultaneous branching of flows that are two flows or above two, while the "XOR template" explains the "mutual exclusion" of those flows that are greater or equals to two. The second Definition of (WF-net) is a Petri Net "N = (P, T, F)" such that: "P" contains a place i with no incoming arcs (the starting point of the process);

- "Sequence template"
- "AND/OR template"
- "XOR template"

which are some basic workflow templates that can be modeled using "Petri Net notation" (Zayoud et al., 2019a). "Petri Nets" (Van der Aalst, 1998), especially a subset known as Workflow (WF) Nets, are the most common process representation used by process mining algorithms like α algorithm (Duda et al., 2012). Petri Nets allow concurrency to be explicitly modelled in a succinct way. Concurrency is a critical feature of business processes, distinguishing them for example from finite grammars, since sequences of activities in different parts of a process may be executed in parallel by different people, perhaps synchronizing at particular points in the process. Petri Nets are executable, with formal semantics, enabling formal analysis of processes. There are various types of Petri Net, details of which including their properties and executable behavior can be found in Van der Aalst (1998). For a discussion of Workflow

Nets, a restriction of Petri Nets commonly used in business processes. The workings of the Petri Net are defined by the marking and the firing of transitions. Transition t may fire when there is a token in each of its input places, where upon a token is removed from each of the input places of t and a token added to each of the output places of t. Places are shown by circles, and tokens by black dots in places. Following the execution of a single transition, more than one following transition may be enabled to fire next (Zayoud et al., 2019a).

In the Figure below a Petri Net representation of the healthcare model that is proposed early in this chapter, it shows the flow of the events in term of places and transitions from start until the end (Zayoud et al., 2019a).

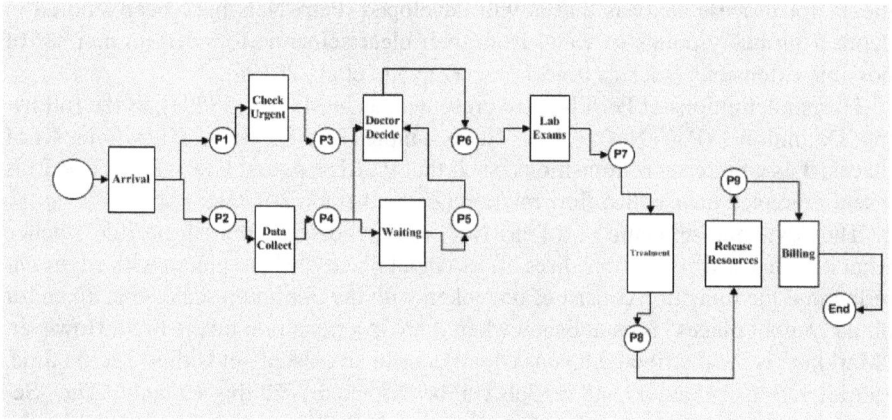

FIGURE 4.4
Petri Net for healthcare example.

4.11   DATA MANIPULATION

The rapid growth of data types force business developers to use new techniques that are able to manage complicated data sets that the traditional applications cannot deal with them in a reliable way. However, there are many factors that make data is considered as big data which has variety of challenges such storing data, data capturing, searching, sharing and analyzing data or any other task that can be implemented on data. Talking about big data, hence the four factors should be defined which are "Volume, Variety, Veracity, and Velocity" (Géczy, 2014). It is important to explain the concept of big data in my book study because of the relation between the new process mining method and its ability to handle different challenges and situations and big data is one of those cases that may appear in rapidly growing organizations. In general "Big data" as a term means to apply predictive, and user behavior analytic to get valuable information from raw data that can be of huge size, and follow many other factors that are explained previously, hence BG is data with process mining together that are applied to discover knowledge and learn methods of organizations with high and advanced scale of data sets weather in size, growing speed, or any other facts that belong to the meaning of big data (Géczy, 2014). The advantages of using process mining methods that can handle big data as well are detecting complicated events, and their correlations which have many benefits especially in healthcare domain such as spotting issues, detecting cures methods, and preventing diseases (Zayoud et al., 2019a).

• Data in Healthcare applications

Data is defined as the features or elements of an organization. In reference to hospital which is the main major in healthcare, data can be the properties of different elements in a hospital such as name, id, age, symptoms, and gender of patients or any other important information for other hospital's elements. The meaning of data has big difference than the meaning of procedures that are named processes which are defined as the way of doing tasks and deploying instructions on the existed resources of an organization (Weijters et al., 2006e). Even though, data and process are representing different methods but they have many common things such as being part of business intelligence and the basic methods for big data analysis in addition to their aim to achieve greater aspects. Both data and processes are implementing particular algorithms that are called mining algorithms some of them are considered data mining algorithms and others are process mining algorithms. However, all mining techniques are aimed to achieve better performance and improved business decisions.

The implemented tools of big data proved its robustness in various domains especially in healthcare because they solve the different issues such as the intervention of clinical risks, reducing waste, reducing unnecessary care services, reporting patient data whether external or internal, making medical terms as standards, and medicine personalizing and analytic of prescriptive (Katal et al., 2013). Some areas of improvement are more inspirational than actually implemented. The level of data generated within healthcare systems is not trivial. With the added adoption of eHealth and wearable technologies the volume of data will continue to increase. This includes electronic health record data, imaging data, patient generated data, sensor data, and other forms of difficult to process data. There is now an even greater need for such environments to pay greater attention to data and information quality. "Big data" very often means "dirty data" and the fraction of data inaccuracies increases with data volume growth. Human abilities are limited when it comes to the scale of big data, and in some cases such as health services the need for intelligent tools is increased with the huge growth of data to ensure accuracy and reliable control of managing information (Zikopoulos et al., 2011a). Moreover, healthcare data is mostly presented in an electronic form that show how much it fits to be considered as big data (Kaymak et al., 2012).

• Processing Data Techniques

There are two main steps in large amount of data' processes which are the data management and analytic. However, data management involves processes and supporting technologies to acquire and store data and to prepare and retrieve it for analysis on the other hand, analytic refers to techniques used to analyze and acquire intelligence from big data. Thus, big data analytic can be viewed as a sub-process in the overall process of "insight extraction" from big data, all these steps are mentioned in the Figure 4.5 (Zayoud et al., 2019a).

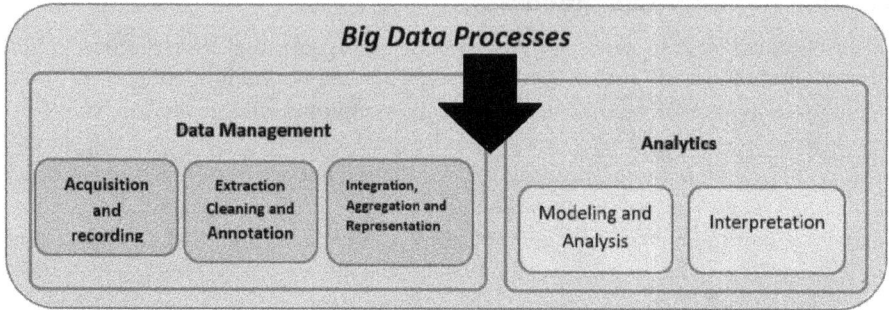

FIGURE 4.5
Processes of analysis using big data
Source: Géczy (2014).

• Big Data Engine "Hadoop"

As seen before, that processing big data is a challenging manner and rational database engines show many limitations in this field. Therefore, new techniques that work based on effective methodologies are implemented and researches are still working on improving these techniques and tools (Zayoud et al., 2019a).

Hadoop is a powerful data engine that can solve different challenges of data especially big amount of data and big data challenges. Hadoop is an open source project hosted by Apache Software Foundation. The idea behind Hadoop, is the distributed computing in which many small sub projects belong to the category of distributed computing infrastructure are deployed. Hadoop basically consists a file system that is called Hadoop File System, and Programming Paradigm or MapReduce. The other sub-projects provide complementary services, or they are building on the core to add higher-level abstractions. The problems of big data are not limited to small challenges, there exist many problems especially in dealing with storage of large amount of data. Even though, the storage capacities of the drives have increased massively, but the rate of reading data from them hasn't shown that considerable improvement. Moreover, the reading process takes considerable amount time and the process of writing is also slower. As a solution, this time can be reduced by reading from multiple disks at once (Géczy, 2014). Only using one hundredth of a disk may seem wasteful. But if there are one hundred data-sets, each of which is one terabyte and providing shared access to them can be considered as a solution. There occur many problems also with using many pieces of hardware as it increases the chances of failure. This can be avoided by creating redundant copies of the same data at different devices so that in case of failure the copy of the data is available. The main problem is of combining the data being read from different devices. Well, many methods are available in distributed computing to handle this problem but still it is quite challenging. All the highlighted and discussed problems are easily handled by Hadoop. The problem of failure is handled by the Hadoop Distributed File System and problem of combining data is handled by Map reduce programming Paradigm (Zayoud et al., 2019a). MapReduce basically reduces the problem of disk reads and writes by providing a programming model dealing in computation with keys and values (Holmes, 2012; Zikopoulos et al., 2011a). Hadoop thus provides a reliable shared storage and analysis system. The storage is provided by HDFS and analysis by MapReduce (Zayoud et al., 2019a).

• The Big data Engine Hadoop VS. Other Techniques

Hadoop is compared with HPC and Grid Computing Tools, the approach in HPC and Grid computing includes the distribution of work across a cluster and they are having a common shared file system. The jobs here are mainly compute intensive and thus it suits well to them unlike as in case of Big data where access to larger volume of data as network bandwidth is the main bottleneck and the compute nodes start becoming idle. Map Reduce component of Hadoop here plays an important role by making use of the Data Locality property where it collocates the data with the compute node itself so that the data access is fast. HPC and Grid Computing basically make use of the API's such as message passing Interface (MPI). Though it provides great control to the user, the user needs to control the mechanism for handling the data flow. On the other hand, Map Reduce operates only at the higher level where the data flow is implicit, and the programmer just thinks in terms of key and value pairs. Coordination of the jobs on large distributed systems is always challenging. Map Reduce handles this problem easily as it is based on shared-nothing architecture, i.e., the tasks are independent of each other. The implementation of Map Reduce itself detects the failed tasks and reschedules them on healthy machines. Thus, the order in which the tasks run hardly matters from programmer's point of view (Holmes, 2012; Zikopoulos et al., 2011b). But in case of MPI, an explicit management of check pointing, and recovery system needs to be done by the program. This gives more control to the programmer but makes them more difficult to write. Moreover, Hadoop is compared with Volunteer Computing Technique. In volunteer computing work is broken down into chunks called work units which are sent on computers across the world to be analyzed. After the completion of the analysis the results are sent back to the server and the client is assigned with another work unit. In order to assure accuracy, each work unit is sent to three different machines and the result is accepted if at least two of them match. This concept of Volunteer Computing makes it look like MapReduce. But there exists a big difference between the two the tasks in case of Volunteer Computing are basically CPU intensive. These tasks makes these tasks suited to be distributed across computers as transfer of work unit time is less than the time required for the computation whereas in case of MapReduce is designed to run jobs that last minutes or hours on trusted, dedicated hardware running in a single data centre with very high aggregate bandwidth interconnects. Last comparison is between Hadoop with RDBMS: The traditional database deals with data size in range of Gigabytes as compared to MapReduce dealing in petabytes. The Scaling in case of MapReduce is linear as compared to that of traditional database. In fact, the RDBMS differs structurally, in updating, and access techniques from MapReduce.

However, creating dimensions of all the data being store is a good practice for big data analytic. It needs to be divided into dimensions and facts. All the dimensions should have durable surrogate keys meaning that these keys can't be changed by any business rule and are assigned in sequence or generated by some hashing algorithm ensuring uniqueness. Expect to integrate structured and unstructured data as all kind of data is a part of big data which needs to be analyzed together. Generality of the technology is needed to deal with different formats of data. Building technology around key value pairs work is recommended (Holmes, 2012; Zikopoulos et al., 2011b). Appendix C shows how the proposed algorithm is implemented using Hadoop concepts to detect events, dependencies, and probabilities of events of big data log file.

As a summary, analyzing data sets including identifying information about individuals or organizations privacy is an issue whose importance particularly to consumers is growing as the value of big data becomes more apparent. Data quality needs to be better. Different tasks like filtering, cleansing, pruning, conforming, matching, joining, and diagnosing should be applied at the earliest touch points possible. There should be certain limits on the scalability of the data stored. Business leaders and IT leaders should work together to yield more business value from the data. Collecting, storing and analysing data comes at a cost. Business leaders will face information technology leaders who must look for many things like technological limitations, staff restrictions etc. The decisions taken should be revised to ensure that the organization is considering the right data to produce insights at any given point of time. Investment in data quality and metadata is also important as it reduces the processing time (Zayoud et al., 2019a).

4.12   BASIC OUTLINES OF β ALGORITHM

The proposed algorithm is as following:

1. β algorithm is written in mathematical representations and formulas while α algorithm not.
2. The math representations of β algorithm makes it applicable in real life application,however it is difficult to apply α algorithm in real life applications.
3. In β algorithm matrix describes topology of events that is called the process.
4. Logical relation between events like XOR, AND, OR, Parallel, and other relations add value on the proposed algorithm.
5. β algorithm is considered as event learning because:
   • Defining the log files help to identify the events (distinct sets of events).

- Defining the dependency metric leads to learn the relations between these distinct events.
- Learning the events from log files can't be complete without including the time and duration's of these events to know the ordering them" (Zayoud et al., 2019a).

TABLE 4.7
Facts about the "β" and "α" Algorithms

α	β
Process includes events	Defines logical operators
Creates workflow nets	Creates Workflow nets
No probability calculations	Calculates the probability of all events
Not written in math representations	Written as mathematical formulas
No frequencies calculations	Calculates the frequencies of events

4.13 CONCLUSION

As a conclusion to this chapter which explains the basic concept of my book. First of all, this chapter starts with highlighting the process mining methods with their most known ones. The outlines are built with a precise survey and a comparison between those methods and their common applications.

The chapter proposed a new algorithm that is called "β algorithm", which is can be considered as process mining algorithm becasue of its ability to learn events from a given log fileevents, their dependencies which are calculated and defined as value "0" for no dependency and value "1" if two or more events are depending on each other. However, the time factor of each event is a critical rule of using this new algorithm,otherwise dependencies are impossible to be calculated. This algorithm is able to know the frequencies of log's events, weather for one event only or events that are corelated with each other with one or more logical relations that are also detected by using the proposed algorithm. Being able to do all the mentioned calculations help to get the probability of an event in a log by knowing the rules of the "probability theory". Moreover, it can calculate probability of events that have one or more of logical relations such as "follow", "OR", "XOR", "AND", "Repeated", or in "Sequence" or "Loop".

To prove the proposed algorithm, a comparison between its results and approach with the most known methods such the α algorithm which can be considered as starting phase of this proposed algorithm named "β".

Petri Nets concepts are deployed to build a Workflow of events that are existed in a log file, hence a workflow net of events in a hospital is built as part of this study. Some metric of the hospital example such as the learned events, dependencies between events, in additon to the frequencies, all of them are presented in this chapter.

Then, the proposed algorithm is tested using "Hadoop" the BG engine, to manage huge amount of data in a log, and stored in different clusters which is the basic methodology of Hadoop to analyze BG. Moreover, an examination of this book's approach is done by designing a platform that can help end users to get good results with positive impact on any organization's performance about its processes and quality assurance. The idea of using BG engine is to optimize the use of my approach to fit any type of organization small, medium, or big ones, especially this innovative approach aims to extract events and anything that can be known about those events to reach the goal of gaining knowledge that will help to build a better version of the current systems' models. The platform is also tested in real life systems that belong to HC by taking surveys form patients and sometimes employees to check the correctness of the gained results of my new approach as I have already published in Zayoud et al. (2019a).

5 Building a Hospital Management Platform

INTRODUCTION

As seen in the previous chapters of this book healthcare organizations are facing the challenge of delivering high-quality services to their patients at affordable costs. For this reason, researchers start to deploy process mining techniques in addition to other available techniques to improve the process in this important domain. The β algorithm that is presented in Chapter 4 of this book, is a new process mining technique that is established to overcome many challenges and issues of the other famous process mining methods such as detecting events from noisy log files, detecting the different logical relations between events, the dependency between those events, their frequency, and probability calculations. The proposed algorithm framework is proved mathematically, and can be applied to handle large sets of data, as the healthcare data.

This chapter, is an implementation of the proposed β algorithm in hospitals and healthcare domain in general as well, hence a new platform called "MYL" is designed based on the β algorithm techniques using some programming skills. "MYL" helps hospitals to implement the mining techniques in practical form to detect the current process model with all its weaknesses and strengths, which provides a clear image of issues in the hospital and opens the way to suggest solutions for deadlocks and problems of the existing processes. Moreover, this new platform can be implemented in many health sectors such as studying the symptoms of diseases and predicting their highest risk factors that cause a particular disease. Hence the "MYL" platform is applied to an Autism disorder to show the benefit of using the platform to increase people's awareness toward this disease and the idea can be extended to any other disease. At the end of this chapter, a survey is provided to patients who visit a hospital at least once per year to validate the output results of the "MYL" platform.

5.1 "MYL" PLATFORM OVERVIEW

The β algorithm is not a theoretical solution only; it is proved mathematically to validate the idea. Then, it is written in an algorithmic format as seen in Chapter 4 of this book. In this chapter, the new proposed β algorithm is implemented as a complete platform name "MYL," which can be used by the end users in hospitals or any other industrial domain.

The idea of this platform is to detect the process model in any organization based on log files that include information about the events in that industrial organization, and their start and end times. This platform is programmed using JAVA programming language which is an object-oriented programming language that gives many

DOI: 10.1201/9781003366577-5

features for the platform such as using many objects classes to identify each part of the organization. Moreover, platforms that are built using Java can deploy graphical user interfaces for easy use of the platform, especially for end users without the need for expert skills. The idea is to predict upcoming events by detecting and understanding the output process model which is need to follow some rules for real-time events prediction. The workflow in hospital based on the proposed algorithm is built based on the extracted data files, which have information about patients and all the phases they go through while they visit the hospitals, after being analyzed, the knowledge discovery starts, with the learning of the events, dependencies, probabilities, and frequencies all to build a workflow model to achieve what is called a predict model that describes the current process clearly according to the flowchart of the β algorithm that is presented in Chapter 4 of this book. "However, the "MYL" platform is built based on the following relations:

1. Occurrence of event.
2. AND relation between two or more events.
3. OR relation between two or more events.
4. XOR relation between two events.
5. Dependency between events.
6. Follow ship between events.
7. Partial parallelism relation between events.
8. Full parallelism relation between events.
9. Sequence, or repeated sequence of events.
10. Finally loop of events" (Zayoud et al., 2019a).

The proposed β algorithm is implemented on huge amount of data that can be considered as big data according to its characteristics. This data is related to big number of patients who visited many hospitals in a specific region, those patients' information are recorded over considerable period of time. However, Appendix B shows the issues when handling it by using traditional rational databases queries such as the MySQL, especially in extracting relations and information, the issue mainly by the limitation of such engines and queries to deal with big data. Hence, the β algorithm is implemented on the same data log files of Appendix C but using big data engine called Hadoop, which shows robustness of the proposed algorithm to manage big data and gives results and statistics. However, the big data concept is not the main concern of my book, hence the proposed method is programmed to build a practical platform that could be used for any industry and this platform as mentioned earlier as "MYL".

After designing this platform, the next phase is to deploy it on real life situations especially the ones that related to the healthcare either for managing hospitals' processes and methods, or for other health purposes such as improving treatments procedures by having the accurate knowledge toward the current issues, or increasing the health awareness in general by analyzing some causes and factors behind some mysterious diseases or disorders.

5.2 APPLYING "MYL" IN EMERGENCY DEPARTMENTS OF A HOSPITAL

To optimize the proposed β algorithm the "MYL" platform is applied to healthcare system example that is related to emergency department in hospital located in Syria and another one in Lebanon, where each country has its own needs regarding the health services, and the system is designed according to these needs. Due to the fact of the different health needs in each country of our world, each system should be studied and analyzed particularly to detect its own problems and apply the required changes and improvements, and this happened by using the right method for that particular system. However, no matter how many differences between healthcare systems, but all health systems are categorized as primary or public healthcare systems, and they have many common factors and issues.

In this section, a study on an emergency department having two different emergency rooms dedicated to public and private sectors in a hospital located in Lebanon is presented, and a comparison between the process model before implementing the β algorithm with "MYL" platform results and after implementing its techniques and its effects on the performance and the entire system is discussed.

"MYL" platform is reading the log files that contain the events and their timing information, in our example it is the patients' records information. After analyzing these log files by the "MYL" which is able to provide the basic events of the system, All the results that output from the "MYL" platform build a knowledge base that is gained accumulatively to learn a process. To clarify the meaning of "learning process" it should be divided into many phases. First, it starts with recognizing events in a log file, the next phase is calculating any dependency that may appear in that log. The third phase begins with calculating probability distributions of those events and their correlation. Building the knowledge of the events, in reference to their correlation, probabilities, and frequencies from those log files, can improve the prediction level of the current process model and workflow net. However, these correlations can be one or more of the relations below:

1. "No relation"
2. "AND" with symbol $\wedge$
3. "OR" with symbol $\vee$
4. "XOR" with symbol $\oplus$
5. "Implication" with symbol $\rightarrow$
6. "Follow-ship" with symbol $\dashrightarrow$
7. "Partial Parallelism" with symbol $\leftrightarrow$
8. "Full Parallelism" with symbol $\Leftrightarrow$
9. "Sequence" with symbol $\mapsto$
10. "Repeated Sequence" with symbol $\curvearrowright$
11. "Loop" with symbol $\circlearrowleft$

All the above relations were explained in detail in Chapter 4 of this book. The coo-relations $\wedge$, $\vee$, $\oplus$, $\rightarrow$, $\dashrightarrow$, $\leftrightarrow$, $\Leftrightarrow$, $\mapsto$, $\curvearrowright$, $\circlearrowleft$ can be clarified as following in reference to hospital example: When two actions occur in a patient in hospital, those actions like arriving at the hospital "AND" getting patient's identity information, and the relation "AND" between those events means to have a condition in term of their execution timing which is forcing one of those events to begin in the mid or during the time of second event's execution. Talking about "OR" relation means any actions or events that belong to this relation should follow the condition of concurrent execution, hence in the example of a patient arriving at a hospital, getting his/her identity, or even the action of checking how much his/her case is urgent all of them can relate to each other by "OR" relation, when one or more of them may be executed. In defining the "XOR" relation, is suitable to mention events like radiology task, blood pressure measurement, and in this situation the "XOR" fits because those two actions logically cannot be executed at the exact time or even having any term of execution time intersection due to many risks that may appear later to nurses and patients as well. As it can be understood from the mentioned clarifications of many logical relations between events, the time of executing events is playing a vital role to define correlation such as "implication, following, partial parallelism, full parallelism, sequence, repeated sequence and loop" between events in any log file (Friedman, 1977b). The relation that is called "implication" is considered when an event's occurrence always follows another event's occurrence such as the "doctor check" event and "arrival to the hospital." On the other hand, the "partial parallelism" happens between "check urgent" and "start treatment" events as an example, where one event begins simultaneously with the second event without having restrictions on the ending period. In "full parallelism," two actions such as "billing" and "discharge from the hospital" begin and finish with same duration. For the "sequence, repeated sequence, and loop," which they mean the appearance of many events and repeated in the same order for different patients, such as the "arrival," "data collection," and "check urgent" events which all of them are considered as sequence relationship. Those events under what is named sequence in case of frequent repetition for various patients in the same log file. Another case is called loop which is a set of events that happen in the same order without any separation such that no distinct events belong to that sequence when those events are repeated frequently in the same order. An example, is in the following relationship: "the doctor checks the patient, the doctor decides, and then patient is moved to radiology, treatment, billing, and finally discharge from the hospital as I have published in article" (Zayoud and Ionescu, 2020; Zayoud et al., 2019a).

The "MYL" platform which is based on the β framework needs a log file or many log files to learn the process accumulatively, and to build an accurate model without the limitations of noisy log files.

• Emergency Room Workflow

A pre-defined model (Oueida et al., 2017) was designed based on site observations, data collected from databases and meetings with medical resources in a hospital that has two main emergency rooms that are named as Emergency room A and Emergency room B.

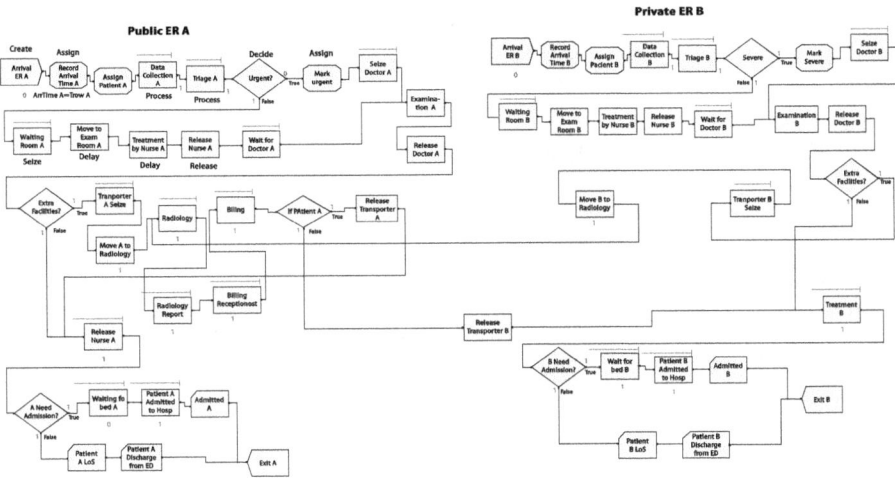

FIGURE 5.1

The Process Model of the Emergency Department

Source: Oueida et al. (2017)

Number In	Average	Half Width	Minimum Average	Maximum Average
Administrator	0.00	0.00	0.00	0.00
Patient	139.50	1.13	135.00	140.00

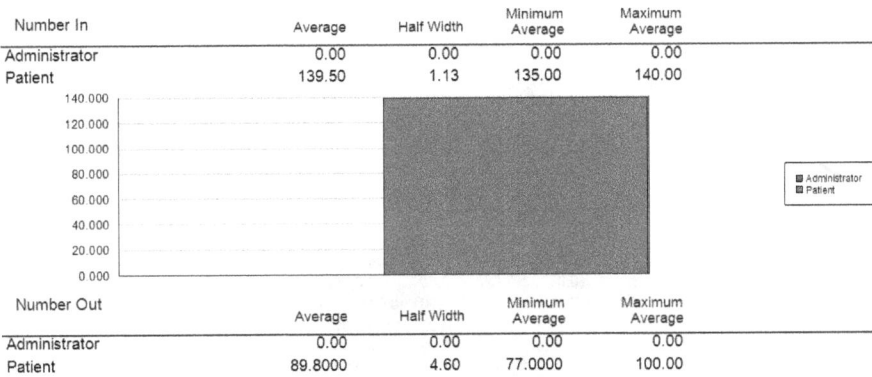

Number Out	Average	Half Width	Minimum Average	Maximum Average
Administrator	0.00	0.00	0.00	0.00
Patient	89.8000	4.60	77.0000	100.00

FIGURE 5.2

Patients Statistics

This model was then simulated using Arena in order to study the system. The following figures describe the model and the statistics collected from simulation (Oueida et al., 2017).

In Figure 5.1, the process model of the current hospital is presented. The emergency department, as discussed earlier, is divided into two different emergency rooms: ER A for public services and ER B for private services. Both emergency rooms (ER) share some resources such as radiology, billing, etc.

Figure 5.2 shows the total number of patients arriving to the ER (Number In) and the total number of patients leaving the system (Number Out).

Waiting Time	Average	Half Width	Minimum Average	Maximum Average	Minimum Value	Maximum Value
Billing Receptionist.Queue	0.06507023	0.06	0.00	0.2661	0.00	5.8073
Billing.Queue	0.00	0.00	0.00	0.00	0.00	0.00
Data Collection A.Queue	0.7392	0.46	0.2459	2.2293	0.00	24.0894
Data Collection B.Queue	0.03955387	0.04	0.00	0.1917	0.00	3.5940
Patient A Admitted to Hosp.Queue	34.2940	4.65	24.0129	43.8236	0.00	88.5155
Patient B Admitted to Hosp.Queue	33.5868	4.21	26.9079	46.1381	0.00	83.7066
Radiology Report.Queue	0.1630	0.09	0.00	0.3638	0.00	18.9679
Radiology.Queue	0.00	0.00	0.00	0.00	0.00	0.00
Seize Doctor A.Queue	0.0994	0.08	0.00	0.2801	0.00	3.3617
Seize Doctor B.Queue	0.0948	0.10	0.00	0.3888	0.00	3.4082
Transporter A.Queue	301.50	41.41	198.96	377.77	0.00	763.80
Transporter B.Queue	326.28	41.51	244.50	395.55	0.00	887.91
Treatment A.Queue	0.1669	0.08	0.04430011	0.3842	0.00	7.6597
Treatment B.Queue	0.07467561	0.06	0.00	0.2658	0.00	4.1617
Triage A.Queue	0.03760471	0.04	0.00	0.1709	0.00	2.7112
Triage B.Queue	0.01709202	0.02	0.00	0.07682275	0.00	3.7450
Wait for bedB.Queue	0.00204241	0.00	0.00	0.02042413	0.00	0.2238
Wait for Doctor A.Queue	0.1383	0.06	0.04685200	0.2547	0.00	5.6521
Wait for Doctor B.Queue	0.1517	0.10	0.04289579	0.5263	0.00	7.4860
waiting for bed A.Queue	0.4134	0.35	0.00	1.2344	0.00	12.6297
Waiting Room A.Queue	0.3726	0.22	0.04392286	0.9753	0.00	26.0936
Waiting Room B.Queue	0.07931656	0.12	0.00	0.5457	0.00	9.5814

FIGURE 5.3

List of Events in the ER

Instantaneous Utilization	Average	Half Width	Minimum Average	Maximum Average	Minimum Value	Maximum Value
Accountant	0.3667	0.01	0.3496	0.3842	0.00	0.5714
Doctor A	0.1880	0.00	0.1785	0.1967	0.00	1.0000
Doctor B	0.1850	0.00	0.1731	0.1953	0.00	1.0000
Nurse A	0.2969	0.03	0.2486	0.4020	0.00	1.0000
Nurse B	0.1959	0.02	0.1579	0.2436	0.00	1.0000
Physician	0.1384	0.01	0.1155	0.1601	0.00	1.0000
Receptionist	0.1562	0.00	0.1458	0.1658	0.00	1.0000
RN A	0.0966	0.00	0.08831071	0.0995	0.00	1.0000
RN B	0.0977	0.00	0.0937	0.1010	0.00	1.0000
Technician	0.07765871	0.00	0.07109614	0.08263649	0.00	1.0000
Transporter A	0.9686	0.02	0.9186	0.9889	0.00	1.0000
Transporter B	0.9600	0.02	0.8971	0.9869	0.00	1.0000

FIGURE 5.4

Medical Resources Utilization Rates

Figure 5.3 lists the events from the studied emergency room, whereas Figure 5.4 presents the utilization rates of the different medical resources serving in this ER. It is clear that transporters are facing the highest workload where their utilization rates exceed 90%. Thus, transporters need urgent attention in order to optimize the process.

Figure 5.5 shows the total time spend by a patient in the system, from the time he arrived at the ER until he/she exits. This total time is referred to as the length of stay (LoS).

Interval	Average	Half Width	Minimum Average	Maximum Average	Minimum Value	Maximum Value
Patient A LoS	254.89	42.91	163.54	363.53	15.6722	790.05
Patient B LoS	268.31	28.32	208.94	326.70	15.1845	928.57

Counter

Count	Average	Half Width	Minimum Average	Maximum Average
Number of Admitted A	11.1000	2.22	7.0000	17.0000
Number of Admitted B	10.7000	2.02	7.0000	15.0000
Patient A Discharge from ED	35.2000	3.44	28.0000	42.0000
Patient B Discharge from ED	32.8000	2.92	28.0000	42.0000

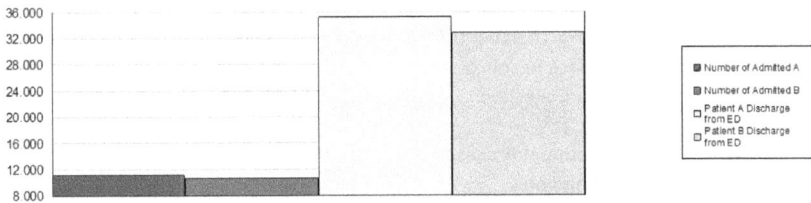

FIGURE 5.5

Patients LoS

The results in the above figures present the process model in the two emergency rooms of a hospital located in Lebanon, where the events and fellowships between those events are collected from around 140 patients, and the model was detected based on the knowledge discovery but using analyzing observations. The current model needed a lot of time to be studied and designed due to the effort needed to observe the events, build a log file and then discover the workflow of that hospital. Some issues facing this system are listed as the following:

▶ Count of patients that enter the system is greater than the count of output patients which cause more deadlocks in case of high demand or catastrophe.
▶ The required time to serve each patient exceeds the accepted average time to meet patients' satisfaction.
▶ The current process shows the need for more resources to serve patients which will increase the cost dramatically, and this is not the aim of this study.

5.3 OPTIMIZING THE EMERGENCY ROOM EXAMPLE USING THE "MYL" PLATFORM

The records of the patients which are shown in Figure 5.6 imported to the "MYL" platform as events log format where each action in the hospital is represented by an event starts at specific time and ends at another time for the same patient ID according to the code derived for the β platform using JAVA programming language.

The dependency and correlations between these events are discovered, then, a similar model as in Figure 5.1 is detected.

This model is discovered by knowing the dependency matrix, correlations and frequency of those events using the β framework automatically after feeding the platform with the data of events as appearing in Figure 5.6.

LogId	Eventname	Start	Ends
0	Billing	1485858180	1484781360
0	Release Resources	1485856620	1484782080
4	Patient Preparation	1483501200	1484775840
5	Doctor Request	1483952760	1484636760
5	Data Collection	1483951800	1484638380
5	Release Resources	1483960140	1484640120
6	ER UNIT	1485723420	1484648940
6	Doctor Request	1485723960	1485094140
6	Discharge Patient	1485732060	1483906140
6	Release Resources	1485731340	1483905300
6	Data Collection	1485809220	1483909800
6	Billing	1485818460	1483911720
6	Treatment Process	1485816240	1485257880
6	Admission	1485814260	1485263520
7	Arrival	1484368560	1485261840
7	Billing	1484378160	1484298780
7	Data Collection	1484369280	1483764720
7	Discharge Patient	1484377560	1483909500
7	ER UNIT	1484740680	1483917120
7	Doctor Request	1484741160	1483912980

FIGURE 5.6

Sample Log File of the Emergency Department

Using the β platform and based on its output results, as shown in Figures 5.7–5.9, the correlations between detected events and their frequencies in addition to their probabilities are discovered. Hence, a knowledge is provided about all the features and problems of the current system.

```
- =========================
- and corelation probability
- =========================
- event names: [admission, arrival, billing, data collection, discharge patient, doctor request, er unit, patient preparation, release resources, treatment p
- event relations: []
- event names frequency: (er unit=5, doctor request=9, arrival=7, release resources=6, discharge patient=6, patient preparation=8, admission=5, data collecti
- and corelation: [admission=doctor request=false, billing=release resources=false, data collection=doctor request=false, data collection=er unit=false, data
- or corelation: []
- xor corelation: [admission=arrival, arrival=treatment process, patient preparation=treatment process]
- partial parallel corelation: []
- full parallel corelation: []
```

FIGURE 5.7

The Events, Correlations, and Their Frequencies

According to the proposed platform, and the knowledge that is extracted, it appears that the dependency between the event named "transporter seize" and "moving to radiology" are related to each other by two relations types: the "AND" and "Partial Parallelism", those two types of relations increased the percentage of the deadlock and added more limitations on the current process model. As a one-tested solution, that does not require any resource addition, therefore, not increasing the cost, is to break the current relations for those two events. This assumption is based on the fact that not all patients need a transporter. For other patients who require some

```
- =============================
- events probability
- er unit   = 0.2631578947368421
- doctor request   = 0.47368421052631576
- arrival   = 0.3684210526315789
- release resources   = 0.3157894736842105
- discharge patient   = 0.3157894736842105
- patient preparation   = 0.42105263157894735
- admission   = 0.2631578947368421
- data collection   = 0.42105263157894735
- billing   = 0.42105263157894735
- treatment process   = 0.15789473684210525
- =============================
```

FIGURE 5.8
Events Probabilities

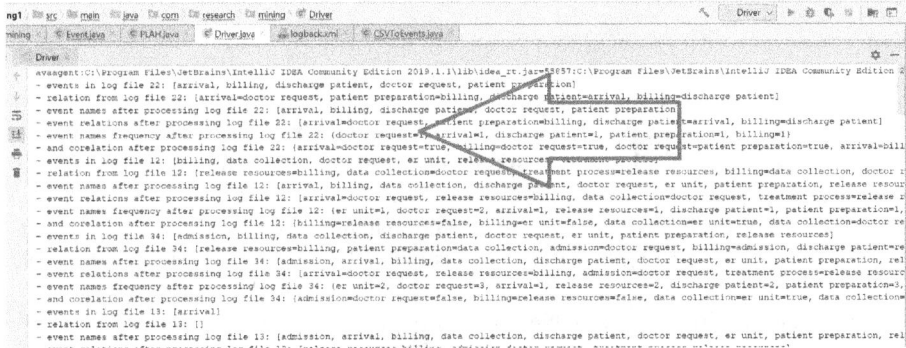

FIGURE 5.9
An Output Sample of the β Platform

medical tests such as radiology images, Citi scans, etc., to accomplish the medical check-up, they can be assisted by a transporter. Record here that the transporter is responsible for transporting patients to the radiology unit when they have difficulties to reach it by themselves. As part of applying the β algorithm, a new decision block is added to the model where the percentage of needed transporter is decreased from 100% to 50% by chance, which means that only patients who are unable to move by themselves to the other unit for accomplishing the extra facilities needed to seize the transporter. This solution will alleviate the bottleneck with the transporter and decreases the need for transporter to seize.

According to the simulation (Alves de Medeiros, 2006) on the same hospital case and under the same features, the output result of this solution shows good improvements on the system: extra patients exiting the system and a lower length of stay (LOS). Recall that no extra resources are added here and thus no extra cost is required.

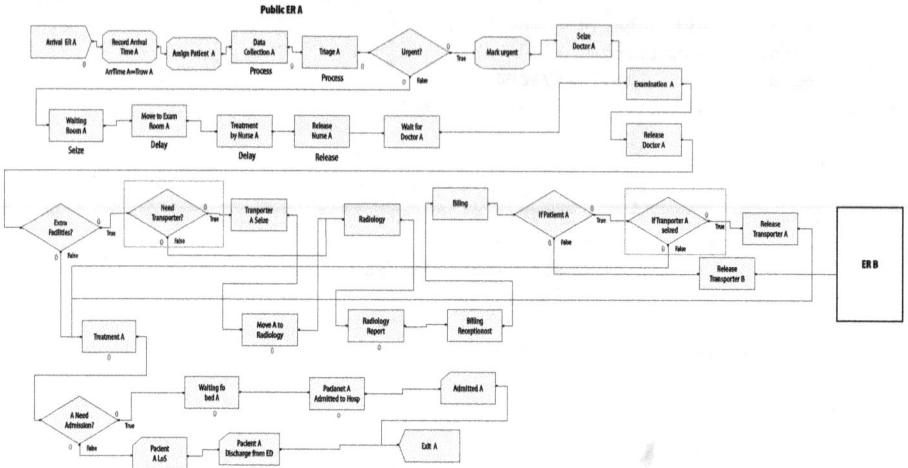

FIGURE 5.10
Arena Simulation for the Modified Process in the Hospital

Simulation results are presented in the figures below where the improvements are noticed, especially, with LoS, and number of patients out. Therefore, an increase in patient and management satisfaction is guaranteed. Patient satisfaction refers to the decrease in the waiting time in the system and management satisfaction refers to the additional number of patients leaving the system and thus a corresponding additional revenue.

The model in Figure 5.10 is the updated model after detecting the events in the system, their correlations, and fellow relationships. A comparison between the model in Figure 5.10 and the original model in Figure 5.1 is presented in Table 5.1.

Running this modified model in Arena for one day with the same simulation settings as of the original model, statistics are collected such as the number of patients out illustrated in Figure 5.11.

Note that the total number of patients exiting the system was 60 patients while here it is increased to 91 patients.

The list of events after modifying the process is presented in Figure 5.12.

Figure 5.13 shows the huge improvement on patients LoS after modifying some of the events behavior without changing dramatically the nature and structure of the current model, especially in ER A.

Figure 5.14 illustrates the percentage of utilization rates after modifying the model. The transporter utilization rates are the point of study where a decrease in the percentage is noticed.

Table 5.2 presents a comparison between the statistics collected from Arena simulation before and after applying the process modifications for ER A and ER B. This comparison shows the robustness and effectiveness of applying such a change and therefore the importance of the β algorithm proposed. According to the results above

TABLE 5.1

Comparison between the Original Model and the Model after Applying β Algorithm

Original Model	Model after Applying β Algorithm
Study and analyze an existed situation discover the relation between all the events and resources in that hospital. Detect the dependency and which of these relations cause the bottleneck in the system. Detect the bottleneck opens the way for many solutions.	Analyze log files of this hospital according to the β algorithm and the "MYL" platform. Detect dependency between transporters and patients' high waiting time during transfer to radiology for X-rays.
Modify the percentage of dependency between transporter and X-ray reach-ability to a level that reduces the average waiting time of the system. As a result, serve more patients in less time.	Increase patient satisfaction by decreasing LoS. Increase hospital management satisfaction by increasing patients Number Out and thus, increasing revenue.

and all the mathematical proofs of this new technique in process mining, it can be concluded that the β algorithm that is presented by the "MYL" platform aims to achieve a better version of any industrial system that needs improvements such as healthcare. This can happen by simply detecting the issues facing his system, suggesting procedures on how to improve it as implemented and labeled by the red arrow in Figure 5.9.

5.4 USING THE "MYL" PLATFORM TO IMPROVE HEALTH AWARENESS

Autism spectrum disorder (ASD) is a developmental disorder (Frith and Happé, 2005). Its symptoms are characterized by impairments in communication with others, and reciprocal social interaction. It is also presented as repetitive, stereotyped behaviors. The clinical picture of autism changed later according to the new researches that show many common symptoms between autism disorder and other disorders such as Asperger syndrome, atypical autism, and disintegrative disorder which are often conceptualized as a spectrum with autism but differs in ages and degrees of functioning (Szatmari, 2003).

The percentage of estimates of the prevalence of autism are 16 per 10,000 before 15 years (Szatmari, 2003), but this estimated prevalence increases to 63 per 10,000 later when all forms of autism spectrum disorders are included. The percentage of this disability is recording increasing percentages from year to another, which is required more serious researches and studies to avoid more cases and help to

	Average	Half Width	Average	Average	Value	Value
Patient	9.2356	0.07	9.0978	9.4071	7.0124	11.7751
Total Time	Average	Half Width	Minimum Average	Maximum Average	Minimum Value	Maximum Value
Patient	168.84	13.81	135.73	202.35	16.2418	947.26

Other

Number In	Average	Half Width	Minimum Average	Maximum Average
Administrator	0.00	0.00	0.00	0.00
Patient	134.00	5.01	119.00	140.00

(Bar chart: y-axis 0.000 to 140.000; legend: Administrator, Patient)

Number Out	Average	Half Width	Minimum Average	Maximum Average
Administrator	0.00	0.00	0.00	0.00
Patient	106.10	5.57	91.0000	117.00

FIGURE 5.11

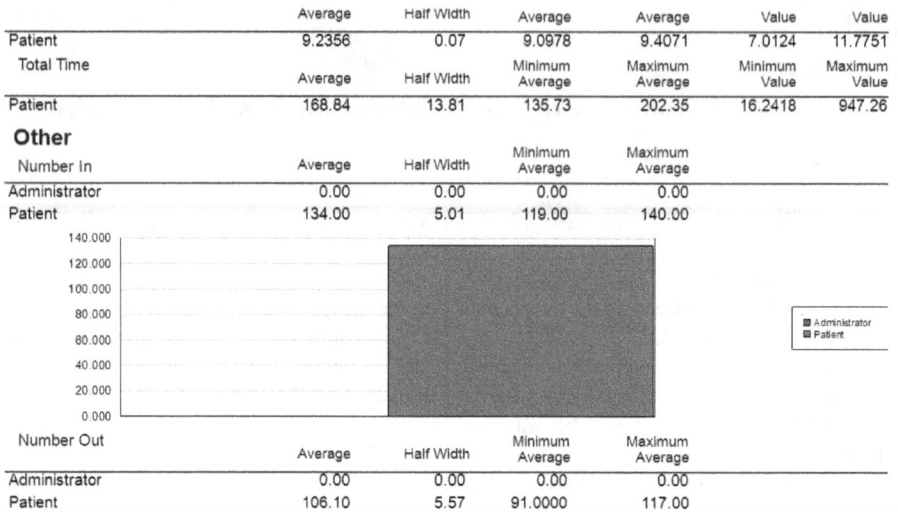

System Number In and Number Out after Modifying the Process

Waiting Time	Average	Half Width	Minimum Average	Maximum Average	Minimum Value	Maximum Value
Billing Receptionist.Queue	0.2689	0.08	0.1126	0.3949	0.00	10.2067
Billing.Queue	0.4913	0.63	0.00	2.8239	0.00	31.6425
Data Collection A.Queue	1.3594	1.09	0.06378692	4.7112	0.00	55.5212
Data Collection B.Queue	0.01516538	0.01	0.00	0.05072343	0.00	3.5506
Patient A Admitted to Hosp.Queue	0.00	0.00	0.00	0.00	0.00	0.00
Patient B Admitted to Hosp.Queue	37.3343	4.81	26.6889	47.8289	0.00	83.8947
Radiology Report.Queue	0.2329	0.19	0.00695530	0.9538	0.00	25.3580
Radiology.Queue	0.00	0.00	0.00	0.00	0.00	0.00
Seize Doctor A.Queue	0.05677266	0.05	0.00	0.1902	0.00	2.3234
Seize Doctor B.Queue	0.06276464	0.06	0.00	0.2395	0.00	3.8318
Transporter B.Queue	298.00	49.99	178.74	404.10	0.00	889.00
Treatment A.Queue	0.1649	0.08	0.02505659	0.3770	0.00	7.1851
Treatment B.Queue	0.0978	0.07	0.00	0.2776	0.00	7.6706
Triage A.Queue	0.01899939	0.02	0.00	0.0977	0.00	3.7021
Triage B.Queue	0.01218728	0.01	0.00	0.03354476	0.00	2.3146
Wait for bedB.Queue	0.03852255	0.06	0.00	0.2137	0.00	2.5644
Wait for Doctor A.Queue	0.06615575	0.05	0.00	0.2431	0.00	4.1987
Wait for Doctor B.Queue	0.0905	0.04	0.03263899	0.1982	0.00	7.0617
waiting for bed A.Queue	0.7933	0.80	0.00	3.7018	0.00	29.6189
Waiting Room A.Queue	0.5608	0.24	0.07090090	1.0423	0.00	30.1660
Waiting Room B.Queue	0.00851759	0.01	0.00	0.02441407	0.00	1.0694

FIGURE 5.12

Events List after Modifying the Process

minimize some of the symptoms. In the early 1980s, researches proved the relations demonstrated the high heritability of ASD and its association with other genetic syndromes (Won et al., 2013), these studies give evidence about the impact of the genetic mutation on the autism as a distinct neurodevelopmental disorder (Won et al., 2013).

User Specified

Tally

Interval	Average	Half Width	Minimum Average	Maximum Average	Minimum Value	Maximum Value
Patient A LoS	89.0079	5.56	76.2228	100.28	16.2418	174.30
Patient B LoS	261.79	48.22	147.45	394.38	16.5258	947.26

Counter

Count	Average	Half Width	Minimum Average	Maximum Average
Number of Admitted A	14.2000	3.21	8.0000	22.0000
Number of Admitted B	10.8000	2.38	5.0000	17.0000
Patient A Discharge from ED	49.4000	3.20	45.0000	57.0000
Patient B Discharge from ED	31.7000	2.26	25.0000	37.0000

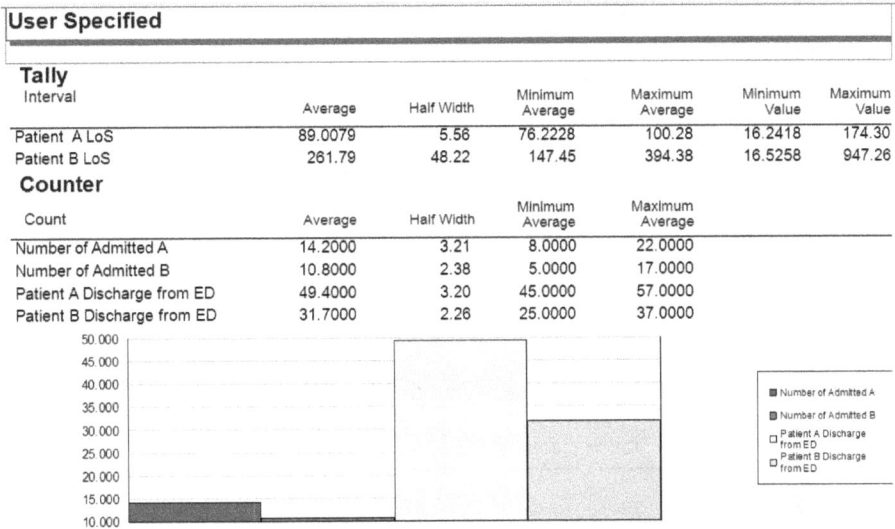

FIGURE 5.13
Patients LoS after Modifying the Process

However, observations from people about their children's behaviors show that many children got the autism disorder after getting the vaccination, MMR. Other observations show the gender of the child has impact on getting affected by this disorder. Moreover, some researches and even treatment methodologies focus on the impacts of the diet nature and the lack of some vitamins during pregnancy on the communication skills of the child in the early stages.

According to doctors and researchers, ASD as disability appears in the early stages of some children, is reduced its symptoms by following specific diet for these children (Won et al., 2013). It is difficult to have clear image about the direct causes of this serious disorder that affects the entire life of the children and families as well. The main problem, is the lack of knowledge which can help to avoid higher percentage values of this disorder. For this reason, technology and many powerful methods that can deal with huge number of information about this disorder, in addition can improve the knowledge base, is required. Process mining algorithms which should be used to improve performance and productivity in many domain, can be applied in this situation as well to provide an accurate model that represents the flow of specific disease/disorder, the method should be probabilistic to analyze the symptoms and the causes as well, hence a clearer image of that disorder will result from the output results. This chapter is defining the process mining in general, process mining in health sector, then how we can implement the process mining techniques in a specific system and then the β algorithm (Szatmari, 2003) is explained to show its impact on discovering the required knowledge for detecting the ASD, and improve awareness toward its disabilities.

Instantaneous Utilization	Average	Half Width	Minimum Average	Maximum Average	Minimum Value	Maximum Value
Accountant	0.4871	0.03	0.4133	0.5448	0.00	1.0000
Doctor A	0.2044	0.01	0.1724	0.2271	0.00	1.0000
Doctor B	0.1765	0.01	0.1578	0.1901	0.00	1.0000
Nurse A	0.3275	0.05	0.2141	0.4378	0.00	1.0000
Nurse B	0.1914	0.02	0.1418	0.2588	0.00	1.0000
Physician	0.1845	0.02	0.1334	0.2355	0.00	1.0000
Receptionist	0.2015	0.01	0.1743	0.2195	0.00	1.0000
RN A	0.0929	0.00	0.08192714	0.0996	0.00	1.0000
RN B	0.0934	0.00	0.08227160	0.0989	0.00	1.0000
Technician	0.1028	0.00	0.08606052	0.1096	0.00	1.0000
Transporter A	0.01444754	0.00	0.00820387	0.02463187	0.00	1.0000
Transporter B	0.9612	0.03	0.8588	0.9877	0.00	1.0000

FIGURE 5.14

Resources Utilization Rates after Modifying the Process

TABLE 5.2

Metrics Comparison before and after Applying β Algorithm

Metrics	Before Applying β Algorithm	After Applying β Algorithm
Number out	60	91
LoS (minutes) for ER A	345.04	74.37
LoS (minutes) for ER B	368.21	131.05
Transporter utilization rate (%) for ER A	98.97	95.27
Transporter utilization rate (%) for ER B	98.92	93.20

5.5   THE IMPACT OF PROCESS MINING FOR BUILDING KNOWLEDGE

As defined earlier in this book that the process mining is the knowledge extracted from event logs which are stored as data files by an information system that is applied in an organization (Weijters and Van der Aalst, 2003). Even though, the information systems help to get the event logs but these logs are rarely used to analyze the underlying processes. Hence, the mining techniques aim to develop, and discover an existed process depending on information given by log of events (Weijters and Van der Aalst, 2003). It is a result of the need of emerging between the data mining (Rebuge and Ferreira, 2012) and business process management (Rebuge and Ferreira, 2012), where data mining is concerned with analyzing the different types of data sets while managing processes in business concerns with designing models of these data types and files. Hence, the process mining plays the middle ware role between the two phases which are "analysis" and "modeling." Knowing that mining processes allows to extract process models from log of events that can identify the existed business processes, and that is done based on information systems that should be implemented in companies. As a result of process mining, a better understanding of activities to force businesses organizations to achieve an optimum procedure than the ones that were implemented before applying process mining methods. However, the output models of process mining algorithms can be graphical for the use of analysts and management teams. Those models can also be formal for the of mathematical analysis. Those models are varying because there are many algorithms under the process mining umbrella, and all have their own features and advantages to discover knowledge and process models types, but each organization or field of study is requiring a suitable method to provide better model and improve the system. The existence of various methods does not mean achieving a solution for each difficulty such as the ones presented in Van der Aalst and van Dongen (2002), and here the most important ones:

- some process models may have the same activity but with the multiple times of appearance, or in various locations. Knowing, that the methods of mining processes cannot provide such tasks most of the time: alternatively, they follow the trial technique which is about adding extra event in the mined model, that causes in most cases wrong connections in the extracted mined model.
- logs reports contain huge amount of data that is not deployed by mining methods, especially data with detailed timing values.
- the accuracy of the output results is affected because of the existed mining methods can not apply a "holistic mining" of various aspects, especially when they are coming from many sources.
- Noise and incompleteness of log files are not always handled by process mining algorithms, and this problem effects the resultant model.

According to the above points, using process mining methods in healthcare is not an easy task, especially if the required knowledge is for building knowledge and workflow of a specific disease or disorder. In this section, the "MYL" platform is applied to provide a better description of the causes of the Autism spectrum disorder (ASD) and increase health awareness toward that disorder (Van Der Aalst, 2011).

• Applying the "MYL" Platform on ASD

A log file that represent data of ASD is provided as input to the "MYL." The autism spectrum disorder (ASD) is broad term used to describe a group of neurodevelopmental disorders (Van der Aalst and van Dongen, 2002). These disorders are characterized by problems with communication and social interaction. Children with ASD often demonstrates restricted, repetitive, and stereotyped interests or patterns of behavior. The scope of this example, is focusing on analyzing this disorder, using the β algorithm and the "MYL" platform which is probabilistic process mining technique as I wrote in Zayoud et al. (2019a), and extracting a knowledge base that improves the awareness regarding this problem. The causes of the ASD are input to the "MYL" platform as log file where each cause represents an event and start means the age where this cause starts to have effect, and the end means when this cause does not have an impact of developing this disorder.

logId	eventName	startTs	endTs
0	A family history of autism	0	1
0	Pregnancy complications, including prematurity.	2	3
2	Advanced parental age	2	2
2	Closely spaced pregnancies	1	2
3	some vaccines, including the MMR	2	2
3	Diet system for the pregnanet mother	3	2
4	Diet system for the baby	2	3
4	Gender Male	4	5
2	A family history of autism	2	3
1	Gender Male	8	9
5	A family history of autism	6	7
2	Diet system for the pregnanet mother	2	2
1	Diet system for the pregnanet mother	3	4
2	A family history of autism	7	8
6	A family history of autism	9	10
7	Diet system for the baby	1	2
8	Gender Male	2	2
1	Gender Male	4	5

FIGURE 5.15
The Input (ASD) Log File Sample

The sample data which includes all the common causes and the start year of their effects are shown in Figure 5.15, and the idea behind studying this sample data, because ASD is not easily identified. The knowledge that is extracted from the "MYL" platform shows the dependencies in some events; which events refer to the causes in this application, their frequencies in different patients, their correlations, and the probabilities of each cause and their correlations for this disorder can produce a model that shows what are the relations between causes and highlights the important ones and gives less attention to the less important ones.

The input files of Figure 5.15 are processed into the "MYL" platform and correlations between the known causes, and their frequencies, all lead to calculate the probabilities of each cause, hence the high-risk factor should be handled with serious action toward it.

The probabilistic results that are presented in Figures 5.16 and 5.17 prove how the genetic mutation that corresponds with this disorder, is highly affected by the family history of this disability. These results, play an important factor in many aspects such as:

1. Informing the parents who have a family history of this disorder about how should manage the other factors carefully before taking the decision of having a baby.

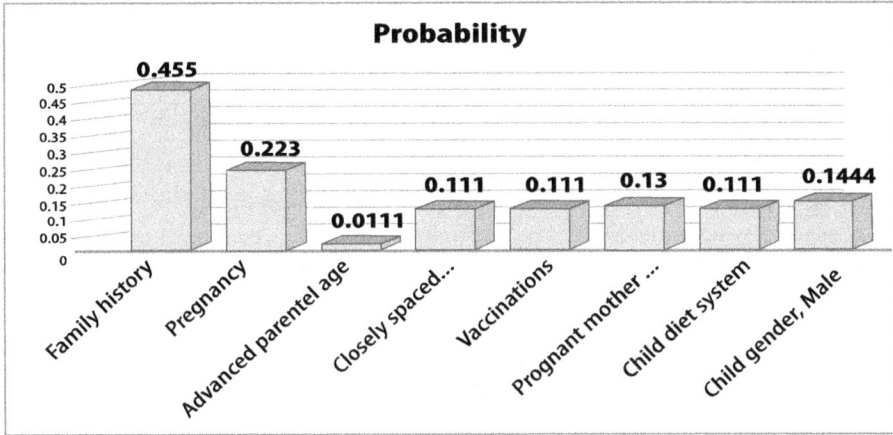

FIGURE 5.16
The Probabilistic Output Result of the ASD Causes

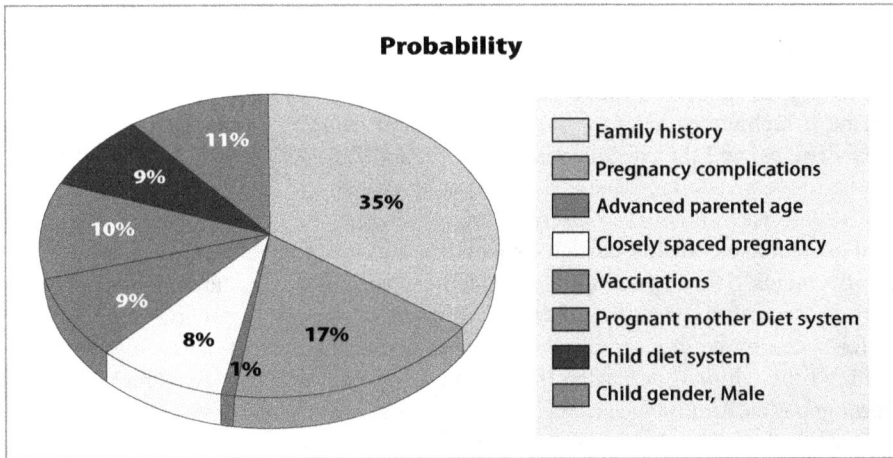

FIGURE 5.17
(ASD) Causes in Another Graphical Representations

2. The knowledge base helps other parents, who have fears about giving the required vaccination for their children, to follow the vaccination schedules if other factors do not exist in their case.

3. It shows the importance of the diet system of mother before even the child born yet.

4. The diet system of the child or the adult who has this disorder, have an impact on the symptoms, and may be reduced in case of following specific diet style.

As a result, analyzing the ASD may have an impact on reducing the percentage of children who will have this problem later, it can manage the symptoms especially in the severe cases after having the clear image of this disorder.

5.6 CONCLUSION

The β algorithm is proved mathematically and implemented on different sets of data types especially data that are extracted from the healthcare sector. The β algorithm is integrated with Hadoop big data engine to handle many challenges which face process mining methods such as the challenge of handling the noisy log files of any organization, or handling large sets of data, where the events and their logical relations are detected in accumulative manner. The β algorithm is implemented and programmed to establish a complete platform called "MYL" based on JAVA programming language.

This "MYL" is able to provide the required knowledge of any system, and detect the process model with all its aspects, weaknesses, and strengths as well. The probabilistic nature of the β algorithm is used to manage the evidences and causes of events in the different kinds of studied systems. The "MYL" platform is optimized by implementing its concepts and techniques on log files, related to a hospital located in Syria and Kuwait as well. The process model was previously designed and studied through observations and site visits. Then, as part of applying the β algorithm and its "MYL" platform, those same log files of that same hospital are analyzed using the β techniques in order to detect all events in the system, their correlations, dependencies, and their frequencies and probabilities. As a result, a process model similar to the pre-defined one is detected automatically.

The automated method of discovering the knowledge of that system provides limitations and issues of the current process, in addition to providing suggestions for improvements. The new scenario is implemented and tested on arena. After running the process with the new modifications suggested, the output reports show a better performance by getting more patients exiting from the system, and a lower patient waiting time without increasing the overall cost or adding extra resources; thus, increasing patient stratification. The above results show the robustness of the proposed platform and the benefits of implementing it in healthcare and in many other industrial domains.

Moreover, "MYL" platform is also deployed to increase health awareness toward the (ASD) in the second application. Its features are applied on the autism spectrum disorders (ASD) to predict a knowledge base that is capable of highlighting the highest risk factors behind this disorder, minimizing the fears toward some other factors, and controlling the disorder symptoms by knowing the correlation between some of the causes, which can improve the health awareness's about the (ASD) and health in general. The study is implemented by choosing a sample of people who suffer from this disorder. Then, an analysis to this sample is provided to show a clear image about this disorder' causes and symptoms.

These results show the highest risk factors that can develop this disorder in the early ages of some children. Hence, gaining the knowledge base improves the awareness of the parents toward many important factors. The awareness of the autism disorder helps to avoid many causes and reduces the percentages of this disorder among the new generation, and helps during the healing stages. Finally, the aim of this chapter is to present the importance of gaining knowledge. This knowledge needs to propagate accumulatively to provide more information that has great impacts on peoples' health in general and can increase awareness or improve health the system as shown in both Zayoud et al. (2019a) and Zayoud et al. (2019d).

6 The Advantages of Using the MYL Platform

INTRODUCTION

The different mining techniques whether data mining, process mining, or any other mining methods aim to improve procedures in industrial organizations which have a great impact to maintain these organizations success and ablity to face the challenges of the tremendous growth of industrial needs. Famous methods such as management information systems (MIS), rational database queries, etc., are not enough to handle all these developments and challenges in business domains that are happening in our world. Moreover, the huge increase of demands that refer to people, industries or societies are required to deploy more powerful techniques and platforms that are able to handle huge amount of information, different types of procedures, and many other challenges and problems. Those innovative platforms and methods can overcome the weaknesses of the other methods and solve the issues in organizations to enhance their performance and maintain them in the market, in addition, they help to meet the different types of customers' demands.

As seen in this book the process mining is the middle-ware stage between data analysis, and business processes management in the organizations. In other words, process mining can be defined as the knowledge extraction practice from log files that contain information about an organization, this information is related mainly to events in the system. The knowledge that is extracted using the process mining methods, is a deep learning of the existed processes in the system with all their issues, weaknesses, and also their strengths. However, the outcome results of mining the process of an organization, is an accurate process model that includes all information about the system procedures, limitations, and powerful points. This extracted business model helps to improve the process and overcomes many deficiencies in that system, hence it increases the overall performance and productivity in that organization. That goal is reached either when developing part of the events or some factors in a given log file; it is also can be achieved by implementing some procedures in that system. As seen in Chapter 3 of this book that process mining has different methods and algorithms, each has its limitations and challenges. Choosing the suitable process mining for an organization is also a challenge to propose a right business model that helps organizations to increase their segmentation by targeting the factors that have a great impact on that organization. Chapter four of this book proposes a new innovative process mining algorithm named the β algorithm and proves its soundness and correctness mathematically and by simulated results. Moreover, in chapter five "MYL" platform is designed using some programming techniques which is a real implementation of the proposed β algorithm. In chapter five the "MYL" platform is applied on two examples related to healthcare system one in hospital and the one related to disease predictions and awareness. Both examples in that chapter show the robustness of using this new platform by the gained results. In this chapter, the robustness of using the "MYL" platform in organizations, especially in healthcare domain is presented by trying this platform in two medical centers for weeks and some questioners in term of survey, are given to the patients of these centers to check their satisfaction after implementing the platform. In addition, SWOT analysis is presented which is a powerful technique in economic studies, to analyze the overall performance and compare the output results of the system after trying the "MYL" platform in a hospitals of the Middle East. The results of the tables in this chapter are filled based on the surveys and some questions of those surveys are presented in this chapter.

6.1   THE ECONOMIC BENEFITS OF USING PROCESS MINING

Using mining methods for processes has an ideal feature, they provide knowledge about the existing process in the system, and how this process works based on evidence and using the log data files that are stored in the management information system of an organization, which allows an objective reconstruction of the process flows (Van Der Aalst, 2012).

The economic benefits of using process mining algorithms are:

• Less time consumption

The process discovery of the activities in a system may need weeks or months, which is not a short time for any project, especially at the beginning of the project where the values of this project are not clear yet to the clients. The ability to implement a process mining brings the first results and hypotheses faster, which helps to return trust and attach people to the remaining part of the project without consuming a considerable time using traditional methods (Van Der Aalst, 2012).

• Understanding the system

The procedures of mining processes present the objectives of building processes it the way they are and gives an answer to some questions such as why people are working in a specific way, and this is rarely detected using the observation or data analysis. However, understanding the root causes of inefficiencies also on the human level is crucial to successfully implementing organizational change, hence it maximizes the value of getting deep knowledge that is normally not discovered (Van Der Aalst, 2012).

• Get a head start within new domains

Process mining helps consultants, who are specialists in a specific domain and are able to provide assistance in their domain only, to understand the new domains also and approach their job by having a tool to understand the process quickly, and increase the productivity, especially of junior analysts, right from the start (Van Der Aalst, 2012).

• Help clients to justify changes within the company

Process mining provides an objective reference for the clients of any organization, which puts them in a stronger position to achieve their goals successfully (Van Der Aalst, 2012).

• Compare before and after

As mentioned before, the process mining understands the process quickly, especially for process improvement projects, where we need to demonstrate the effect of the implemented changes. For example, showing how the process has been streamlined after a change (and one month of new data collection) by comparing the "before" and "after" images. Ideally, the process performs much better now, and uses process mining to communicate these results (Van Der Aalst, 2012). The above benefits show the impact of using the process mining techniques on any organization, especially by reducing the time of analysis, the cost of understanding the process, and getting the objective references of clients' goals, those benefits represent important part of the economic factors.

6.2   QUALITY IMPROVEMENT OF THE MEDICAL SERVICE

SWOT analysis is an efficient method to measure the quality of any service, since the focus of my book is on healthcare sector hence SWOT analysis is considered a powerful procedure to measure the quality of the provided medical services. However, SWOT analysis is a strategic managerial technique used by organizations and individuals in order to identify the strengths, weaknesses, opportunities, and threats related to existing competitions and strategic planning. The paradigms of strengths and weakness are closely related to the internal factors of businesses while threats and opportunities are recognized as the external factor. Figure 6.1 shows the four dimensions of SWOT analysis, namely:
1. Strengths: are characteristics and skills that gives the company/individual competitive advantage over other players in the market.
2. Weaknesses: are characteristics or shortcomings of a company/individual that makes that brings a disadvantage relative to other players.
3. Opportunities: are elements in the environment which are giving profitable opportunity in the industry and can be exploited to company's advantage.
4. Threats: are elements in the environment that could become an obstacle for the company and industry's operation.

• Benefits

These internal and external factors are used to generate an effective plan expressed as Strategic fit. SWOT presents decision-makers and managers with an easy and retainable form of information. It is also used to check the feasibility and practicality of an investment. When it comes to medical sciences there is no margin of error one

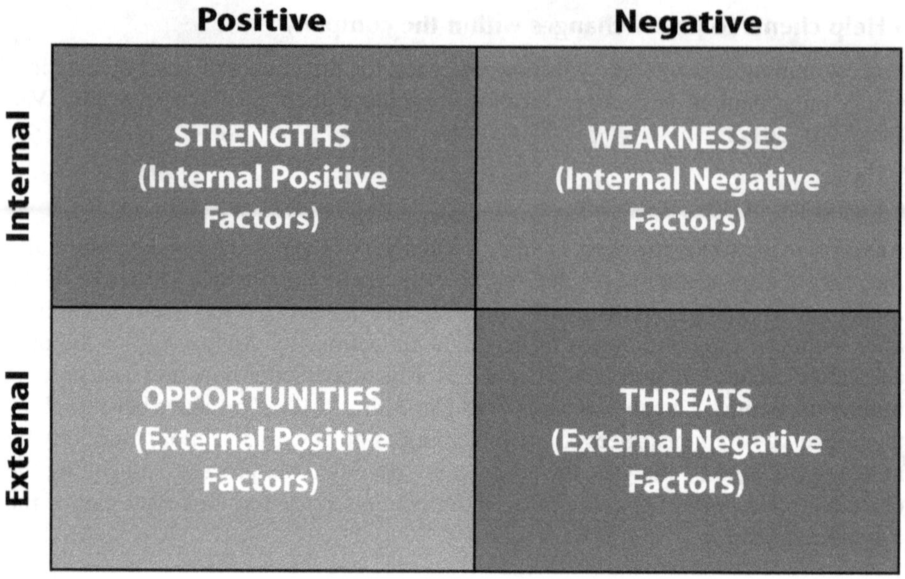

	Positive	**Negative**
Internal	STRENGTHS (Internal Positive Factors)	WEAKNESSES (Internal Negative Factors)
External	OPPORTUNITIES (External Positive Factors)	THREATS (External Negative Factors)

FIGURE 6.1
SWOT Analysis Means
Source: Sabbaghi and Vaidyanathan (2004).

must be 100% correct all the time. Even the most trivial errors can result in fatal consequences. Tragedies due to human error are very common in the medical industry. Today, by deploying powerful algorithms, doctors are able to better diagnose and execute surgeries through lasers and surgical instruments controlled through computers. These measures improve the accuracy and efficiency of the process while minimizing the collateral damage. Those algorithms have made the way we treat patients more effectively.

• Strengths

The medical industry collects and generates large amount of raw data that must be mined and sorted. Every year almost a million medical studies are published. The human brain is brilliant but there is a limit to the amount of information that it can process and store. The β algorithm is one of the algorithms it can be applied to all aspects of our everyday life. In mathematical terms, it is functions, that are used for the learning processes, and it is able to handle complex material into simpler components using big data engine concepts.

• Simplicity and accuracy

The β algorithm is a powerful technique and may be implemented to help clinicians make incredibly accurate determinations about our health from large amounts of information, premised on largely unexplainable correlations in that data. With extraordinary accuracy, this algorithm is able to predict and diagnose diseases, from cardiovascular illnesses to cancer, and predict related things such as the likelihood of death, the length of hospital stay, and the chance of hospital readmission. All these advantages are because it is a learning technique and probabilistic method.

• Weakness

Weakness like strength are inherited characteristics of an organization or program, but instead seen as the problem's weaknesses are usually evaluated as the areas of improvement. Although using algorithms in medical practices and diagnosis can improve the accuracy and efficiency of the tasks being performed but with the great amount of advancement comes along its disadvantages too.

First, using an algorithm is a time-consuming practice if there is no friendly-user platform. It takes hours and even days of man hours in order to develop a code or program that can run with specific given conditions. Second, the development of algorithms takes up a considerable time in adjustment and its modification is required along.

Last, applying the β algorithm as steps and procedures needs an educated and skillful personnel for development and operation such a system which not only bring the cost of hiring an individual but also brings in high maintenance cost which increases the operating leverage of medical colleges and institutes. Hence, designing the MYL platform as a tool of the β algorithm was a must to gain benefits and good results for industry and improve all the decision-making variables to an exponential level.

• Complexity and sustainability

Cloud computing and massively parallel processing systems have created hope to implementation of these techniques for short term, but as data volumes go up, and deep learning moves toward automated creation of increasingly complex algorithms such as the β algorithm.

• Opportunity

Opportunities are openings or chances for something positive to happen. They usually arise from situations outside your organization and require an eye to what might happen in the future. They might arise as developments in the market you serve, or in the technology you use. Being able to spot and exploit opportunities can make a huge difference to your organization's ability to compete and take the lead in your market.

• Optimization

Using the β algorithm in diagnosis, open a whole new area of optimized digital diagnosis which allows patients and doctors to use a huge connected cloud network that can connect them to global resources used for diagnosis and treatment. It opens opportunity for medical students to learn and interact with professionals to generate insights about the medical world and receive counseling regarding their career counseling. Last, a number of times it has been observed that in critical cases, which doctors prefer to discuss the case with their fellow expert to bring another thinking direction to the subject using the β algorithm connects all these brilliant minds behind a code to analyze the cause-and-effect relationship accordingly.

• Threats

Threats are the external factors that make the sustainability of a business or process difficult in long run. However, the β algorithm has a very promising future in terms on medical businesses, but it also holds potential threats in the industry, especially in medical practices because it is a delicate matter regarding holding the legal bounding and incident or accident that may result in severe consequence hence machine dependency is still not favored by many medical management professionals. Moreover, SWOT analysis can be used to analyze the economic factors of an organization, and it shows the benefits to the industry of using the process mining methods, by studying the strengths, weaknesses opportunities, and threats that may face any organization from implementing the specific method. A SWOT analysis is often used at the start or as part of a strategic planning exercise, and process mining affects the strategic decisions and directs them to the suitable operational actions at the right time. Nowadays, organizations face many challenges and threats that cannot be handled by using one method, hence merging between process mining methods results and evaluating these results by the SWOT analysis, provide efficient study of the system, suggest better solution, and take a powerful step at the right time without making dramatically lose or changes in that organization. SWOT analysis provides the consultants with an economical vision of the system output model that is an extracted result of applying the mining techniques, in addition, SWOT analysis evaluates the mining method and its results by comparing the system's strengths, weaknesses before and after applying that mining algorithm.

However, the proposed MYL platform is deployed on two medical centers located in Kuwait which is presented as the first medical center according to the table name, a nd the second one in Syria which is presented as second medical name, and two different surveys were given to sample of patients after implementing this platform in that medical center to check the improved levels and compare the current system model after the new modifications. The questioners are listed in this chapter with their results and each question shows the percentage of patients' who responds. The questions are managed to identify the internal factors of success of implementing the new techniques and the external factors for managers in that healthcare centers. The keys of success are: reducing the time of service in the care providers, increasing the overall patients' satisfactions, improving the quality of provided health services and managing the cost to be reasonable and not increased. However, Tables 6.1 and 6.2 is showing those internal factors of success of applying the MYY in healthcare provider place and their weigh, score according to survey, and their values that are calculated according to the formula: Weigh × Score and the result should greater than 1 to have the confident about the new system that is applied, knowing that the value of any factor is an indicator to know which factors need to be improved in the future.

TABLE 6.1
Analysis of Internal Factors for Success for the First Medical Center

No.	Internal factors	Score	Weigh	Value
1	QOS	8	0.4	3.2
2	LOS(Time)	7	0.2	1.4
3	PS	7	0.3	2.1
4	Cost	5	0.1	0.5
	Total	27	1	7.2

TABLE 6.2
Analysis of Internal Factors for Success for the Second Medical Center

No.	Internal factors	Score	Weigh	Value
1	QOS	9	0.4	3.6
2	LOS(Time)	6	0.2	1.2
3	PS	8	0.3	2.4
4	Cost	4	0.1	0.4
	Total	27	1	7.6

The scores are extracted from the survey, and each number in the table is out of 10. The QOS means quality of service, LOS and time of service, PS patient satisfaction. The weigh is distributed according to importance of each factor in that particular health center, and the values show that the first three factors are improved while the cost is need to be improved for the next phase in the future.

Moreover, the external factors that affect the success of an organization, and they are presented in Tables 6.3 and 6.4, which follows the same criteria as Table 6.1. However, the external factors such as improving the medical services in reference to the treatment procedures (TP), integration with more intelligent procedures (AI), handling organization that has data considered as big data (BD), and finally having log files that include important information related to the time of each activity (time).

TABLE 6.3
Analysis of External Factors for Success for the First Medical Center

No.	External factors	Score	Weigh	Value
1	TP	6	0.4	2.4
2	AI	7	0.2	1.4
3	BD	4	0.2	0.8
4	Time	4	0.2	0.8
	Total	21	1	5.4

TABLE 6.4
Analysis of External Factors for Success for the Second Medical Center

No.	External factors	Score	Weigh	Value
1	TP	7	0.4	2.8
2	AI	8	0.2	1.6
3	BD	3	0.2	0.6
4	Time	5	0.2	1.0
	Total	23	1.0	6.0

According to the result in Table 6.3 the main threats for implanting any new system in healthcare provider are having information that is growing rapidly and categorized as big data, in addition to another threat which is the completeness of the log file of activities.

Table 6.5 is filled with all the basic strong points, weak points, opportunities and the threats of using the MYL platform, which are all the points of the SWOT analysis.

6.3 ANALYSIS OF PATIENT SATISFACTION

The new framework "MYL" represents a new solution for managing existing data over an existing process model, and to prove that this platform and the whole solution is valid hence it is implemented on mainly two hospitals one located in Kuwait State and the other one in Syria because simulating and running the current solution using only simulators and mathematical approaches are not enough and leads to try and deploy the new solution in real environments which are in our situation should be hospitals and medical centers. The real implementation in hospitals needs to be

TABLE 6.5
SWOT Analysis of MYL Platform

S	W
1. Easy to Use	1. Information in log files should be organized according to specific criteria
2. Handle noisy log files	2. Big data concepts are required when MYL needs to be implemented on industry that has big data.
3. Probabilistic approach	
4. Applicable with big data domains	
O	**T**
1. Good for improving treatments in the future by implementing its prediction methods	1. Log files without timing information make the results, not accurate.
2. Can be integrated with aritificial intellegent concepts to deliver smart healthcare systems.	2. Big data engines are must in case of dealing with rapid growing industries.

validated by giving the patients questions that can provide clear feedback about the results and advantages and disadvantages of using the new solution. Therefore, two surveys which their questions and their results are presented in this section to show how much the new solution "MYL" improves the satisfaction of patients that are visiting those hospitals after applying the MYL, The following are the main points that are discussed in the given surveys:

- Time needed to set an appointment.
- Time needed to serve patients.
- Time needed to get treatment.
- Quality of the provided services.
- Organization of the whole process.

All the above points got very good results according to patients responds in the given surveys which emphasizes the soundness of the MYL and the β algorithm as well in addition to their advantages in improving the whole system. The following figures are examples of the two surveys about the patients' opinion and level of their satisfactions about important factors in the medical center after modifying the system according to MYL platform results and recommendations.

The survey questions are as follows:

The provider means the medical center which provides health services.

Q1.

Overall, how satisfied or dissatisfied were you with your last visit to our office?

Answered: 31   Skipped: 0

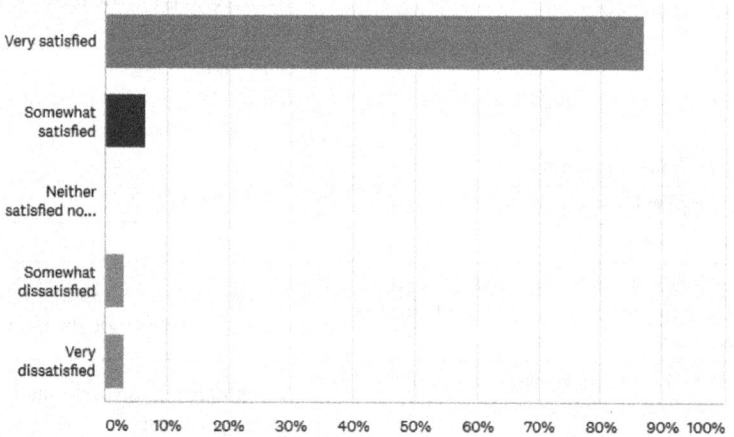

ANSWER CHOICES	RESPONSES
Very satisfied	87.10%
Somewhat satisfied	6.45%
Neither satisfied nor dissatisfied	0.00%
Somewhat dissatisfied	3.23%
Very dissatisfied	3.23%

The above question shows that more than 87% of patients were satisfied after their last visit after implementing the new changes resulted from MYL platform.

Q2.

How easy or difficult was it to schedule your appointment at a time that was convenient for you?

Answered: 31   Skipped: 0

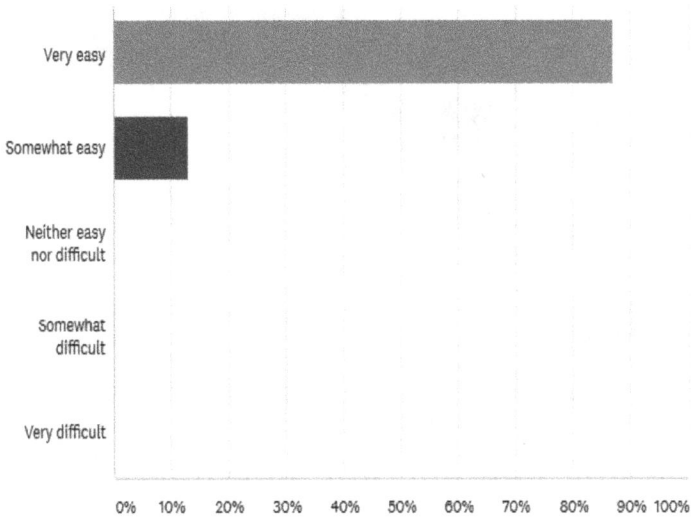

ANSWER CHOICES	RESPONSES
Very easy	87.10%
Somewhat easy	12.90%
Neither easy nor difficult	0.00%
Somewhat difficult	0.00%
Very difficult	0.00%

The second question shows the ease of getting appointment by getting high score from the patients opinions.

Q3.

How convenient was the appointment time you were able to get?

Answered: 31   Skipped: 0

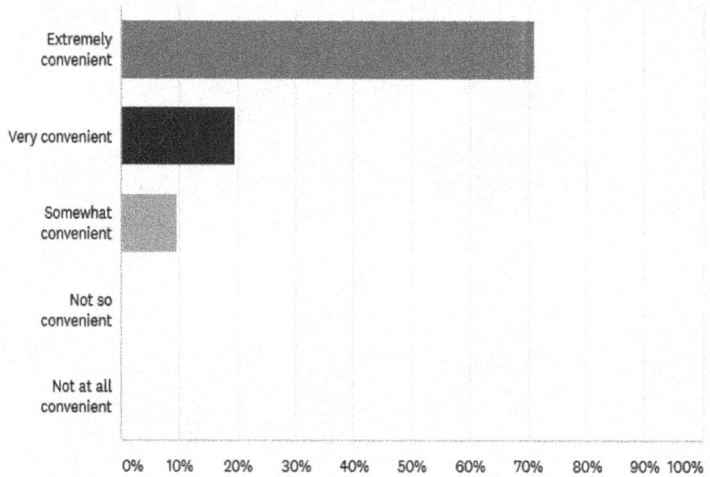

ANSWER CHOICES	RESPONSES
Extremely convenient	70.97%
Very convenient	19.35%
Somewhat convenient	9.68%
Not so convenient	0.00%
Not at all convenient	0.00%

Q4.

In your opinion, how convenient is the location of our medical center

Answered: 31   Skipped: 0

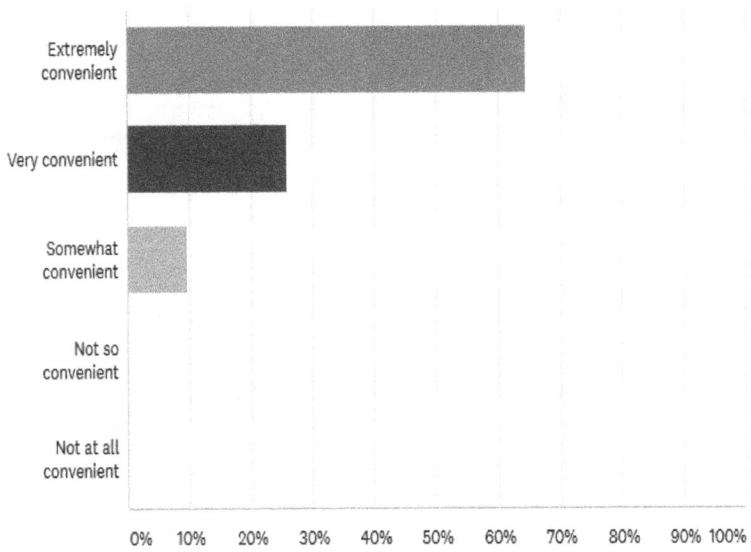

ANSWER CHOICES	RESPONSES
Extremely convenient	64.52%
Very convenient	25.81%
Somewhat convenient	9.68%
Not so convenient	0.00%
Not at all convenient	0.00%

Q5.

Overall, how would you rate the service you received from the staff ?

Answered: 31   Skipped: 0

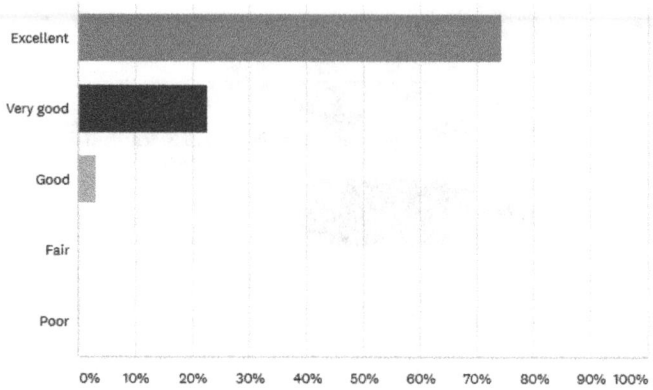

ANSWER CHOICES	RESPONSES
Excellent	74.19%
Very good	22.58%
Good	3.23%
Fair	0.00%
Poor	0.00%

Q6.

Did your appointment with your provider start early, late or on time?

Answered: 31   Skipped: 0

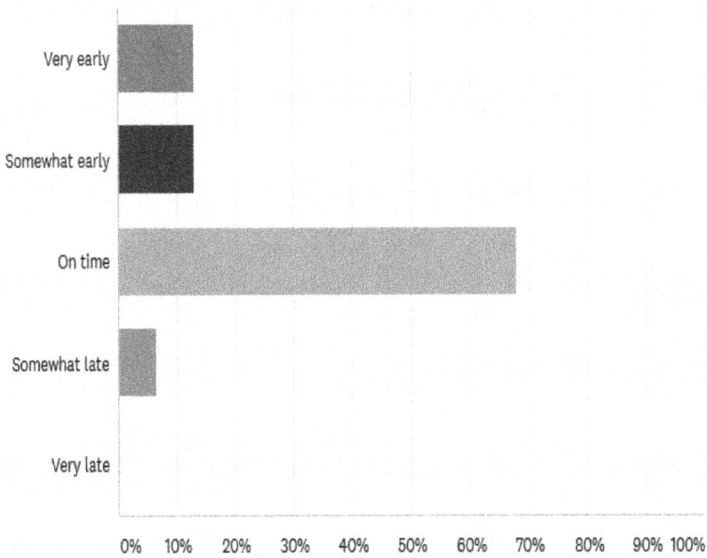

ANSWER CHOICES	RESPONSES
Very early	12.90%
Somewhat early	12.90%
On time	67.74%
Somewhat late	6.45%
Very late	0.00%

Overall, how would you rate the care you received from your provider?

Answered: 31   Skipped: 0

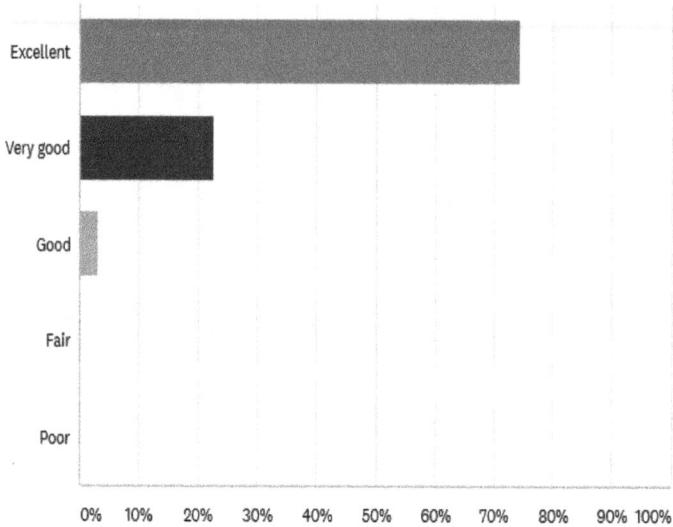

ANSWER CHOICES	RESPONSES	
Excellent	74.19%	23
Very good	22.58%	7
Good	3.23%	1
Fair	0.00%	0
Poor	0.00%	0

Q7.

How much do you trust your provider to make medical decisions that are in your best interests?

Answered: 31   Skipped: 0

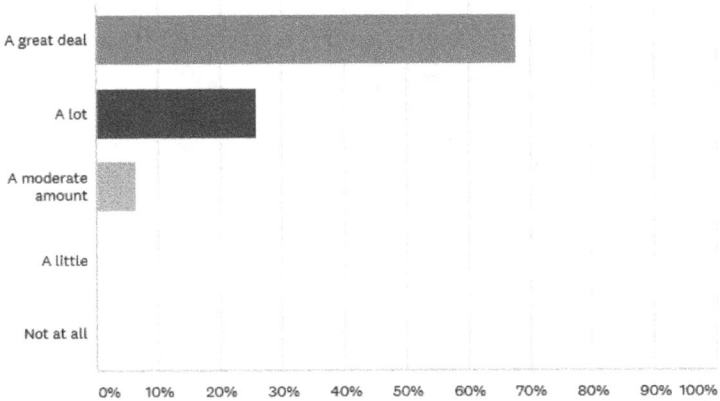

ANSWER CHOICES	RESPONSES
A great deal	67.74%
A lot	25.81%
A moderate amount	6.45%
A little	0.00%
Not at all	0.00%

Q8.

How satisfied or dissatisfied were you with the amount of time your provider spent with you addressing your needs?

Answered: 31   Skipped: 0

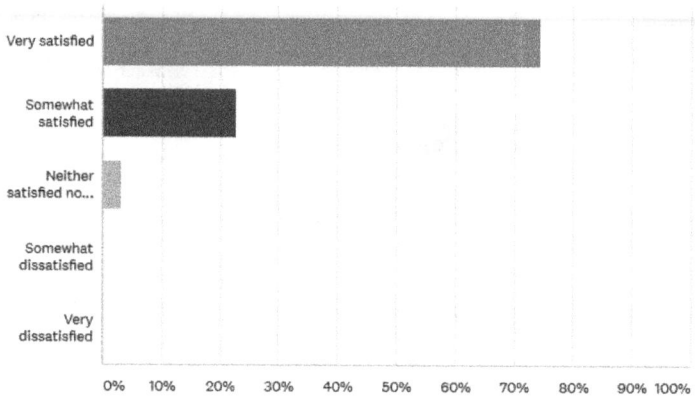

ANSWER CHOICES	RESPONSES
Very satisfied	74.19%
Somewhat satisfied	22.58%
Neither satisfied nor dissatisfied	3.23%
Somewhat dissatisfied	0.00%
Very dissatisfied	0.00%

Q9.

How likely is it that you would recommend your provider to a friend or family member?

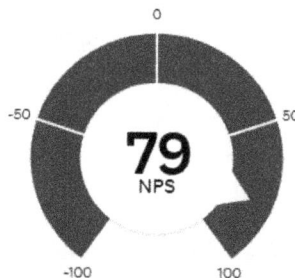

Questions of Survey part two:

Q10.

9. Do you agree that applying an intelligent medical system can improve the health care sector?

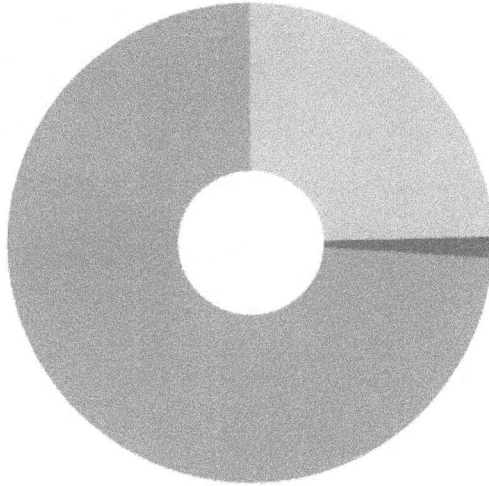

Skipped: 18   Answered: 69

Yes		74%
No		1%
Not necessary		25%

Q11.

8. Do you have confidence in human knowledge?

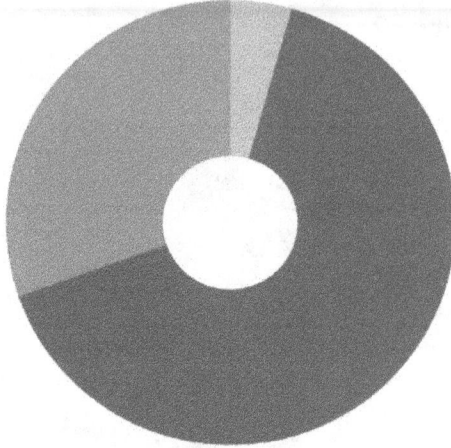

Skipped: 18   Answered: 69

Yes		30%
Human make mistakes		65%
No		4%

Q12.

7. Do you think that sometime the required tests from doctor are not matching your symptoms?

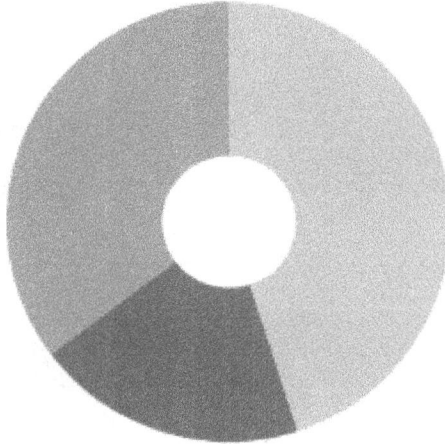

Skipped: 16   Answered: 71

Yes	35%
No	20%
Rarely, but it happened	45%

Q13.

6. Admission to hospital depends on

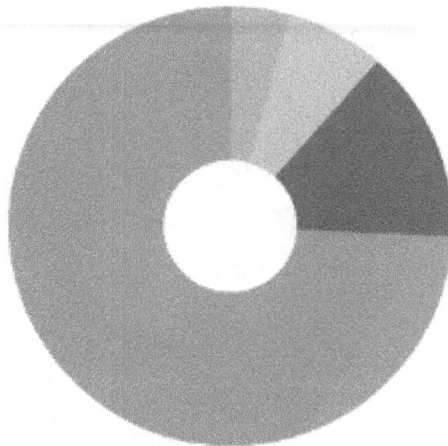

Skipped: 17   Answered: 70

Doctor's decision and tests results		74%
If I have medical insurance		14%
Some symptoms appear		7%
Patient wants to be admitted		4%

Q14.

5. Do you know which spcialest you need to take appointment with for your case?

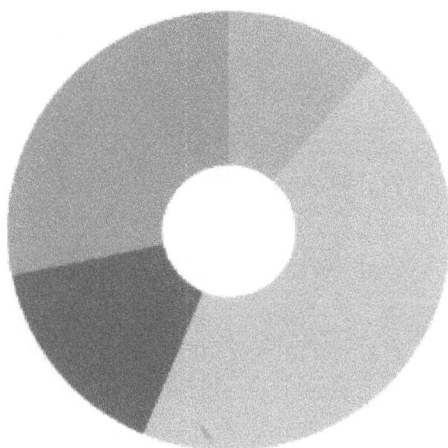

Skipped: 16   Answered: 71

Yes	28%
No	15%
I wait until a general health doctor to decide	45%
I visit many spcialests until I know my problem	11%

Q15.

4. The laboratory tests are taking place upon:

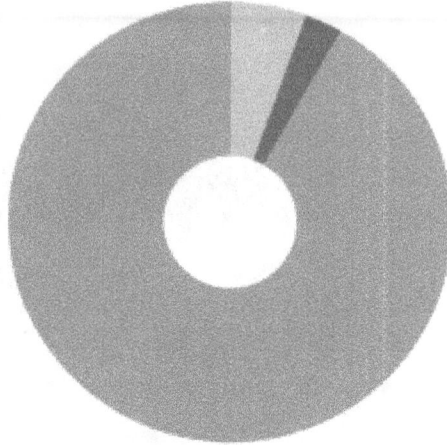

Skipped: 14   Answered: 73

Doctor decision		92%
Nurse decision		3%
Immediately after arrival		5%

Q16.

2. If you have an emergency situation, so what is the action taken from the hospital ?

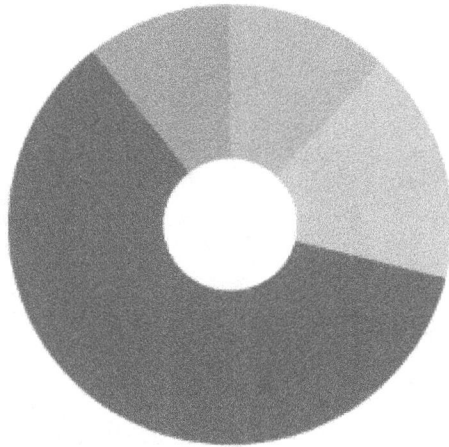

Skipped: 11   Answered: 76

Let you wait so long	11%
Transfer you to ER immediately	61%
Specialist doctor checks you	17%
Do the billing process	12%

Q17.

1. What happened immediately after arriving to hospital ?

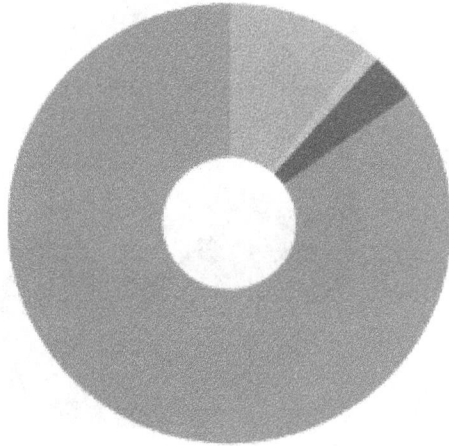

Skipped: 0   Answered: 87

Collecting your information		85%
Start treatment		3%
Xrays		1%
Billing		10%

The main points of the above surveys are:

1. Time needed to set an appointment.
2. Time needed to serve patients.
3. Time needed to get treatment.
4. Quality of the provided services.
5. Organization of the whole process

All the above points give an idea about the improvement of the provided medical services, and show how the healthcare organizations are working hard to increase patients' satisfactions and by making their healthcare experience is with good quality where their experience of care is critical and based on the level of that experience, health organizations can continue or lose their place in the community and the business domain. However, hospitals and healthcare providers use surveys to measure the patients' satisfaction and their experience and this is what was done for the MYL platform based on common factors as noted in the five points above. Moreover, those surveys are valid methods to improve systems and show the level of satisfaction. The surveys that are done and presented in this chapter, collect patient experience data to enable objective, meaningful comparisons between the procedures in hospital before applying the new platform and after it has been implemented for few weeks in the same hospitals. Since survey is an efficient way and due to the results of the two given surveys of the MYL platform, hence MYL platform plays an important role to improve healthcare systems and also patients' satisfaction as well.

6.4   CONCLUSION

This chapter targets the benefits of using the various methods of mining processes in any organization and focuses on the β algorithm which is a new process mining algorithm. This chapter emphasizes on choosing the right method for an organization which is considered as a challenge nowadays because applying the right method gives the required knowledge to improve an organization's performance and highlights the weaknesses and issues to be resolved. A SWOT analysis is defined briefly with its economic benefits and their common goals with the process mining methods and applied on the MYL platform that is a practical implementation of the β algorithm to show its economical value, especially in medical sciences and domains.

In summary, MYL framework shows good results after it has been tried for some weeks in two medical centers and the results are proven by the patients' opinions and satisfaction levels. These results prove the importance of process mining methods and especially the proposed solution to be implemented in industrial domains which face issues and challenges nowadays. SWOT analysis is required to improve business performance because the old traditional methods do not have the ability to protect any organization from failures, hence merging powerful methods like process mining and SWOT analysis techniques is inevitable for continuous improvement and tracking of sustainable cost savings and revenue enhancements in the industry.

The Conclusion and Limitations

Innovation and improvements are logical results for the rapid growth of demands in our world. Industries with complicated systems that contain multi units, groups, resources, and many different activities, are highly required for new techniques and methods to maintain their business and success. This applies to healthcare because it is a huge field in industry, and it affects different aspects in each country and society. Healthcare is a combination of different categories such as patients, specialists, medicines, equipment's, procedures, and many other interconnected facilities which require multi-paradigm, simulation modeling, technologies, and methodologies to handle all these requirements. In this book, healthcare system is chosen to implement new proposed process mining technique on it for the purpose of improving and developing healthcare systems regarding different perspectives such as improving the business models, health procedures, and resource optimizations. The proposed technique in this book is called the "β algorithm" and is an extension of a famous process mining method, this new technique overcomes many limitations in the previous methods, and it is proved to be correct and sound according to the mathematical and simulations techniques. It is developed as a software called **"MYL"** that is implemented in two hospitals for validation and optimization purposes and the results show improvements after the good feedback. However, the "MYL" software is designed as following:

- First, studying the different available methods to improve industry and its processes such as management information systems techniques, and different mining techniques.
- Second, comparing and analyzing those different techniques show the limitations and strengths of each technique.
- Third, providing a survey about these techniques which helps to choose the best method for industry especially in healthcare domain.
- Fourth, process mining is chosen as the base of this book because it helps to handle the existed system's procedures without destroying this system.
- Fifth, a basic and famous process mining algorithm called the α algorithm is chosen specifically because if its simplicity and other advantages.
- Sixth, the chosen algorithm is analyzed and studied to highlight its advantages and disadvantages all to propose a new algorithm called the β which is a better version of the α algorithm and can be implemented on real applications.
- Seventh, the proposed β algorithm is proved mathematically that it is sound and correct regarding improving procedures and detecting issues in any system and able to solve problems and give valid solutions.
- Eighth, the β algorithm is implemented on different health examples, one for hospitals models, and one for analyzing and predicting health issues in some

diseases such as autism disorder, the simulations results of the β algorithm show good improvements in different health aspects.

- Ninth, this new algorithm is designed as real-life software called MYL to be implemented in real-life healthcare systems.
- Tenth, MYL tried by two hospitals and it shows good improvements especially from the patients' and managers' satisfactions.

This book came up with a new solution to improve the management information systems in hospitals that are laid under the umbrella of healthcare domains, the new technique and solution improves by being tested theoretically using well known simulators and practically by being implemented in hospitals which are located in different countries such as Syria, Lebanon, and Kuwait; the following points:

- the quality of service in hospitals,
- the patient's satisfactions,
- reducing waiting and service time,
- and the manager's satisfactions.

However, the surveyed outcomes measure the results and output and give good responses, and the different types of testing such as the mathematical proofs, Arena simulation results, and the output of the software and the survey all provide a good feedback about the new technique and solution that is named "MYL" based on the new algorithm named "β" algorithm which is written in an algorithmic format and proved its soundness mathematically, and then the platform named "MYL" that is implemented based on the β algorithm and tested on real-life examples, one in hospitals and another one for predicting high-risk factors of one mysterious disorder. In those applications, the "MYL" show robustness in gaining knowledge that can understand the current situation with all its strengths and weaknesses and gives ways for improvements and building better situations. This new method and platform are mainly inherited from process mining techniques which are hot topic nowadays in industrial domains to detect issues and provide solutions. **MYL** is applied to a predefined model of emergency department in a hospital located in Lebanon, and after gaining the results as outputs of this software, data are fed to Arena Simulator and the resultant model is modified according to the results of this software where it detects the main cause of deadlock in the system and the highest dependency values between events the lead to problems in the performance. After, applying the modification on the model using Arena simulator and compares the results from the original model and the modified model, the results regarding the number of patients out from the system is increased, and the LOS is decreased for each patient without affecting the total cost, and the overall performance is improved which of course will increase the patients and mangers satisfactions. Not only this example, but the new software is also applied on the Autism disorder and after reading log files; which are the symptoms of specific age and probable causes of this disorder; of the sample of patients that suffer from this disorder, the results from the MYL software presents the correlations between some factors and having this disorder and show the highest risk factors among other which helps to improve awareness of this disorder to minimize the percentage of children that may get this disease. Moreover, the output results of MYL helps to minimize the fears of some actions such as some vaccinations that may cause this disorder by showing the real value of the correlation between this action (giving vaccinations) and the disorder. As a future work, The MYL software is able to be integrated with artificial intelligence concepts to be implemented for improving treatments processes such as improving a prediction system that can help to figure out the main reason for a specific disease after handling big data of statistics and medical cases. This will help to avoid such diseases or improve the curing procedures.

LIMITATIONS

Some limitations to complete this book are illustrated as follows:

- Mainly two hospitals located in Syria and Kuwait were chosen to apply the new techniques on.
- Data collection phase was a difficult phase of this book because the chosen hospitals are located in countries such as Syria, and Lebanon and they still depend on paper-based approaches for data management and storage more than automated information system, and collecting data in this situation was not an easy task.
- however, the new method is applied also in the dermatology department in a medical center located in Kuwait which has an automated information system but it is also not fully automated hence collecting data may not lead to accurate or complete log files of information.
- health information is considered critical data because of the patient's privacy.
- simulating of validity of the β algorithm was done using Arena simulator, while there is also an other simulator that can be used to prove the robustness of the new technique.
- to construct the platform, especially from mathematical perspective only Petri Nets used while there are other types of networks that can be tried
- time for studying and analyzing the economic benefits and patients satisfaction of using the "MYL" software was limited because improving any system need time in months and years to show improvements in reference to the economy, and measuring these benefits need many surveys and questioners in addition to statistics on long term.

A The β Algorithms

This appendix is presenting the algorithms of my solution, and the first algorithm is the one for learning events of a log file:

Algorithm 1 Calculate Λ and $\mathbb{D}$

Require: $\mathbb{L}$
Ensure: Λ and $\mathbb{D}$
  Define $\Lambda_C = \emptyset$
  if Learned before **then**
    Load Λ
    Load $\mathbb{D}$
  else
    $\Lambda = \emptyset$
    $\mathbb{D} = \{0\}$
  end if
  while not end of $\mathbb{L}$ **do**
    Read λ
    if $\lambda \notin \Lambda$ **then**
      $\Lambda = \Lambda \cup \lambda$
      $\|\Lambda\| = \|\Lambda\| + 1$
    end if
    if $\lambda \notin \Lambda_C$ **then**
      $\Lambda_C = \Lambda_C \cup \lambda$
    end if
    for all $\lambda_i \in \Lambda_C$ **do**
      for all $\lambda_j \in \Lambda_C \wedge 0 \le j < i$ **do**
        $\mathbb{D}_C(i,j) = 1$
      end for
    end for
    for $0 \le i < \|\Lambda\|$ **do**
      for $0 \le j < \|\Lambda_C\|$ **do**
        $\mathbb{D}(i,j) = \mathbb{D}(i,j) \times \mathbb{D}_C(i,j)$
      end for
    end for
  end while
  Save Λ and $\mathbb{D}$.

Algorithm 2 calculates "$R\theta$, $V\theta$, and $F\theta$. The input is the log file L, dependency matrix D and the set of learned events ˙ and the output is $R\theta$, $V\theta$, and $F\theta$, where θ is any relation" (Zayoud et al., 2019a).

DOI: 10.1201/9781003366577-A

Algorithm 2 Calculate $\mathbb{R}_\theta$, $\mathbb{V}_\theta$ and $\mathbb{F}^\theta$

Require: $\mathbb{L}$, $\mathbb{D}$ and λ
Ensure: $\mathbb{R}_\theta$, $\mathbb{V}_\theta$ and $\mathbb{F}^\theta$
  for all $\theta \in \{\wedge, \vee, \oplus, \to, \dashrightarrow, \leftrightarrow, \Leftrightarrow, \mapsto,$ **do**
    Define $\mathbb{R}_\theta = \emptyset$
    Define $\mathbb{F}^\theta = \emptyset$
  end for
  for all $\lambda_i \in \Lambda$ **do**
    for all $\lambda_j \in \Lambda$ **do**
      if $(t_j \le t_i \le t_j + \tau_j) \vee (t_i \le t_j \le$ **then**
        $\theta = \wedge$
      else if $(t_i < t_j \vee t_i > t_j + \tau_j) \vee ($ $\vee\, t_j > t_i + \tau_i)$ **then**
        $\theta = \oplus$
      else if $(t_j \ge t_i + \tau_i)$ **then**
        $\theta = \to$
      else if $t_j = t_i$ **then**
        $\theta = \leftrightarrow$
      else if $(t_j = t_i) \wedge (\tau_i = \tau_j)$ **then**
        $\theta = \Leftrightarrow$
      else if $(t_j \le t_i) \vee (t_i \le t_j)$ **then** ——
        $\theta = \vee$
      else if $(\lambda_i \to \lambda_j) \vee \exists \lambda_k \| \lambda_i \to \lambda_k \wedge \lambda_k \to \lambda_j$ **then**
        $\theta = \mapsto$
      else if $\|(\lambda_i \mapsto \lambda_j)\| \ge 2 \wedge \exists \lambda_k \notin (\lambda_i \mapsto \lambda_j) \| t_i(\lambda_i \mapsto \lambda_j) \le t_k \le t_j(\lambda_i \mapsto \lambda_j)$
      then
        $\theta = \curvearrowright$
      else if $\|(\lambda_i \mapsto \lambda_j)\| \ge 2 \wedge \nexists \lambda_k \notin (\lambda_i \mapsto \lambda_j) \| t_i(\lambda_i \mapsto \lambda_j) \le t_k \le t_j(\lambda_i)$
      then
        $\theta = \circlearrowleft$
      end if
      if $\theta \in \{\wedge, \oplus, \to, \leftrightarrow, \Leftrightarrow\}$ **then**
        $\mathbb{R}_\theta(\lambda_j, \lambda_i) = \mathbb{R}_\theta(\lambda_j, \lambda_i) \times 1$
        if $\mathbb{R}_\theta(\lambda_j, \lambda_i) = \ge 1$ **then**
          $\mathbb{F}^\theta(\lambda_j, \lambda_i, c) = (\lambda_j, \lambda_i, c + 1)$
        else
          $\mathbb{R}_\theta(\lambda_j, \lambda_i) = 0$
          $\mathbb{F}^\theta(\lambda_j, \lambda_i, c) = 0$
        end if
      end if
      if $\theta \in \{\vee, \mapsto, \curvearrowright, \circlearrowleft\}$ **then**
        $\mathbb{R}_\theta(\lambda_j, \lambda_i) = 1$
        $\mathbb{F}^\theta(\lambda_j, \lambda_i, c) = (\lambda_j, \lambda_i, c + 1)$ 27
      else
        $\mathbb{R}_\theta(\lambda_j, \lambda_i) = 0$
      end if
    end for
  end for

B Rational Database

The MS SQL server database engine is used to build the database tables according to the above model as shown in Figure B.1, then the data records of all patients are imported as chunks of data due to big amount of records that cannot be imported by one query and one time to the rational database engine. Figure 2.2 shows the tables of database is classified according to the model in Figure 2.1:

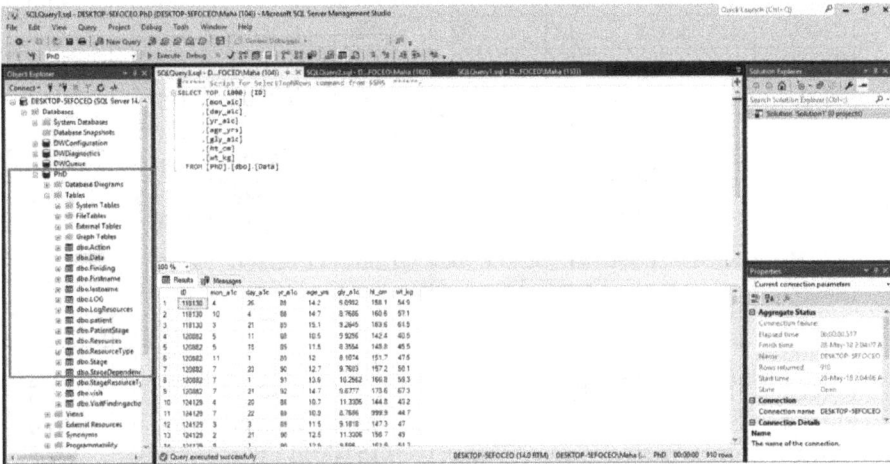

FIGURE B.1

TABLE B.1

Sample of Patients' Information

NUMID	Month	DAY	YEAR	AGE	GLY	HGT	WGT
109140	5	16	86	13.2	10.992	168.1	59.9
108230	11	14	78	11.7	6.7886	160.6	57.1
135130	3	21	89	15.1	9.2645	163.6	61.5
130552	5	11	83	9.45	8.7256	152.4	60.5
130432	5	15	88	11.5	8.3554	148.8	45.5
124592	1	1	79	12	8.1074	151.7	47.5
100782	3	23	60	12.7	9.7603	157.2	50.1
134582	3	1	81	13.6	10.2562	166.8	59.3
123562	3	21	62	14.7	9.6777	173.6	67.3
134569	4	20	78	10.7	11.3306	144.8	43.2
145699	7	22	78	10.9	8.7686	999.9	44.7
100919	3	3	79	11.5	9.1818	147.3	47
123459	2	21	90	12.5	11.3306	158.7	49
134909	6	11	81	15.7	9.595	161.9	51.2
138879	2	22	91	12.4	10.4215	166.4	57.4
134899	9	22	91	13	10.7521	170.2	60.3
124129	2	10	92	14.4	9.3471	170.2	64.5
124129	6	20	92	14	8.6677	999.9	70.9
126139	8	8	88	12	9.0165	158	67.7
145139	4	21	89	10.7	11.4959	165.4	69.5
134569	4	18	89	13	10.0909	168.5	73.2
16789	12	28	89	13.3	10.5868	170	77.4
126139	3	5	90	13.5	10.0909	171.4	78.8
126180	4	19	88	12.6	10.9174	147.8	38.6
13456	6	21	88	12.8	6.9504	148.3	39.3
126180	4	6	89	13.6	7.6942	154.4	45.6
126180	11	28	89	14.2	9.6777	160.3	50.6
126180	4	19	90	14.6	7.6942	163.7	54
126180	10	30	90	15.2	8.686	165.1	56.2
129511	3	5	87	10.3	10.4215	139.6	36.3
129511	9	22	87	10.8	9.595	141.7	39
129511	2	16	88	11.2	9.595	144.9	42
129511	7	26	88	11.7	8.8512	146.5	44.7
124551	11	17	88	12	9.9256	148.1	46.5
134561	2	16	89	12.2	8.8512	148.5	47.7
123456	10	26	89	12.9	9.3471	153.5	52.5
234781	2	22	90	13.2	10.5041	155.9	54
187537	7	12	95	14.4	11.3	171.8	60.9
187537	8	14	95	14.5	9.8	172.3	62.1

Another example from another table:

```
/****** Script for SelectTopNRows command from SSMS
    ******/
SELECT TOP (100) [VisitID]
,[PatientID]
,[VisitNo]
FROM [PhD].[dbo].[visit]
```

TABLE B.2

Sample of 100 Patients' Visits Records

VisitID	PatientID	VisitNo
14874677	1	1
14874678	2	1
14874679	3	1
14874680	4	1
14874681	5	1
14874682	6	1
14874683	7	1
14874684	8	1
14874685	9	1
14874686	10	1
14874687	11	1
14874688	12	1
14874689	13	1
14874690	14	1
14874691	15	1
14874692	16	1
14874693	17	1
14874694	18	1
14874695	19	1
14874696	20	1
14874697	21	1
14874698	22	1
14874699	23	1
14874700	24	1
14874701	25	1
14874702	26	1
14874703	27	1
14874704	28	1

(Continued)

TABLE B.2

(Continued). **Sample of 100 Patients' Visits Records**

VisitID	PatientID	VisitNo
14874705	29	1
14874706	30	1
14874707	31	1
14874708	32	1
14874709	33	1
14874710	34	1
14874711	35	1
14874712	36	1
14874713	37	1
14874714	38	1
14874715	39	1
14874716	40	1
14874717	41	1
14874718	42	1
14874719	43	1
14874720	44	1
14874721	45	1
14874722	46	1
14874723	47	1
14874724	48	1
14874725	49	1
14874726	50	1
14874727	51	1
14874728	52	1
14874729	53	1
14874730	54	1
14874731	55	1
14874732	56	1
14874733	57	1
14874734	58	1
14874735	59	1
14874736	60	1
14874737	61	1
14874738	62	1
14874739	63	1
14874740	64	1
14874741	65	1
14874742	66	1

(Continued)

TABLE B.2
(Continued). **Sample of 100 Patients' Visits Records**

VisitID	PatientID	VisitNo
14874743	67	1
14874744	68	1
14874745	69	1
14874746	70	1
14874747	71	1
14874748	72	1
14874749	73	1
14874750	74	1
14874751	75	1
14874752	76	1
14874753	77	1
14874754	78	1
14874755	79	1
14874756	80	1
14874757	81	1
14874758	82	1
14874759	83	1
14874760	84	1
14874761	85	1
14874762	86	1
14874763	87	1
14874764	88	1
14874765	89	1
14874766	90	1
14874767	91	1
14874768	92	1
14874769	93	1
14874770	94	1
14874771	95	1
14874772	96	1
14874773	97	1
14874774	98	1
14874775	99	1
14874776	100	1

MS SQL server is used for the mentioned records and by experimental tests healthcare data records that have been collected are classified as big data.

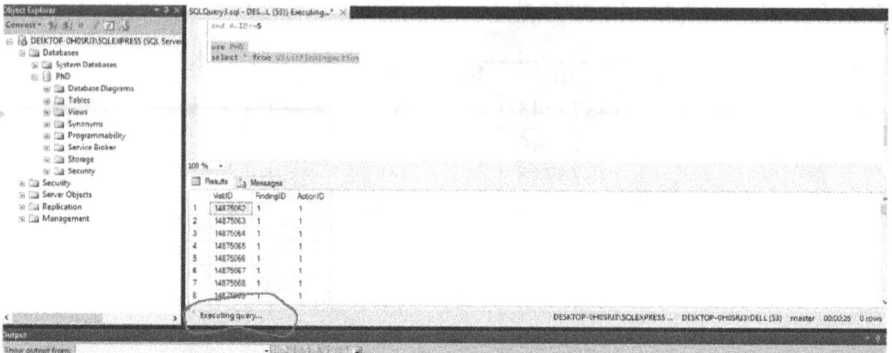

FIGURE B.2
Select Query – starting to execute

Experiment is the best method to prove anything, so I have imported all the mentioned records to MS SQL server after setting the database correctly according to the model in Figure B.1 and the extended α algorithm proposed in chapter one of this report, and during the process the following issues appear:

1. Data records are huge and where cannot be imported by single query method, hence the records are imported into chunks.
2. The time spent to import the records is a considerable factor.
3. The Execution of queries is limited for any method that should be done on the whole records and leads to errors in execution as shown in Figure B.3.
4. Calculating many statistical information is impossible using the rational database engines.

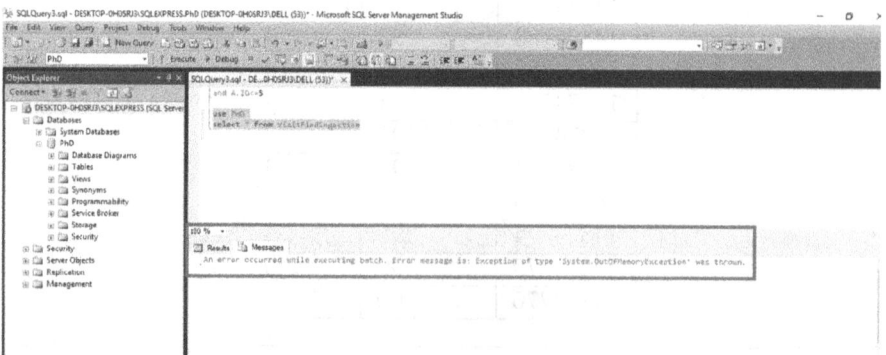

FIGURE B.3
Select Query – after waiting some time SQL server error

All the mentioned difficulties and according to the characteristics of big data and their definitions provide the health care data records that I have, the description of big data.

These limitations and difficulties that appear when handling big data on relational database engines lead us to think about another approach or technology to handle these sets of data in order to mine them and their processes as well. Nowadays, there are many companies and business industries in deep need of big data engines Hadoop, Knime, IBM big data engine.

C The β Algorithm Using Hadoop Big Data Concept

The following figures show how the β algorithm is tested and programmed on Hadoop big data engine by JAVA SPARK programming language to detect the events from any log file of big data, their dependencies, and their probabilities and using the cluster-based concept of big data log files. One part to detect events

```java
private static Iterable<Tuple2<String, Long>> sortCaseDetails(Iterable<Tuple2<String, Long>> caseDetails) {
    List<Tuple2<String, Long>> sortedCaseDetails = new ArrayList<>();
    caseDetails.forEach(tuple -> sortedCaseDetails.add(tuple));
    Collections.sort(sortedCaseDetails,
        (Tuple2<String, Long> o1, Tuple2<String, Long> o2) -> o1._2().compareTo(o2._2()));
    //System.out.println(caseDetailsSorted);
    return sortedCaseDetails;
}

public static void main(String[] args) {
    if (args.length != 1) {
        System.out.println("pass HDFS or local file path to process");
        System.exit(1);
    }
    // args[0] HDFS or local file path to process data
    PLAH.process(args[0]);
}
```

then another part to detect dependencies between events in a log file.

```java
// Dependency learning
// Group by logFileId and sort the values in temporal order
// shuffle :set(LogFileId, (Event, TimeStamp)) - (LogFileId, set(Event, TimeStamp))
JavaPairRDD<String, Iterable<Tuple2<String, Long>>> traces =
    logEvents.groupByKey().mapValues(caseDetails -> sortCaseDetails(caseDetails));
//traces.foreach(tuple -> System.out.println(tuple));

JavaPairRDD<Tuple2<String, String>, Boolean> dependencyLearning = traces
    .mapValues(sortCaseDetails -> generateDependencyInATrace(sortCaseDetails)).values() JavaRDD<Object>
    .flatMap(it -> it.iterator()) JavaRDD<Object>
    .mapToPair(tuple -> new Tuple2<Tuple2<String, String>, Boolean>(tuple._1(), tuple._2()))
    JavaPairRDD<Tuple2<String, String>, Boolean>
    .distinct() JavaPairRDD<Tuple2<String, String>, Boolean>
    .reduceByKey(new Function2<Boolean, Boolean, Boolean>() {
        @Override
        public Boolean call(Boolean v1, Boolean v2) throws Exception {
            return v1 && v2;
        }
    });
//dependencyLearning.foreach(o -> System.out.println(o));
dependencyLearning.coalesce(1).saveAsTextFile( path: "F:\\maha\\OneDrive\\Documents\\output\\dependencyLearning");
```

and finally, the probability calculations of the events.

```
// Probability calculation
JavaRDD<String> eventIdRDD = logEvents.values().map(tuple -> tuple._1());
long totalEvents = eventIdRDD.count();
JavaPairRDD<String, Float> eventProbabilityRDD =
        eventIdRDD.mapToPair(eventId -> new Tuple2<String, Long>(eventId, 1L))
            .reduceByKey((a, b) -> a + b).mapValues(count -> 1.0f * count / totalEvents);
//eventProbabilityRDD.foreach(o -> System.out.println(o));
eventProbabilityRDD.coalesce(1).saveAsTextFile( path: "/tmp/output/eventProbability");
```

D An Overview of the MYL Framework's Source Code

The β algorithm implemented in JAVA programming language:

```java
package com.research.mining;

import javafx.util.Pair;
import lombok.NoArgsConstructor;
import org.slf4j.Logger;
import org.slf4j.LoggerFactory;

import java.util.Collections;
import java.util.HashMap;
import java.util.HashSet;
import java.util.Iterator;
import java.util.List;
import java.util.Map;
import java.util.Set;
import java.util.TreeSet;

@NoArgsConstructor
public class PLAH {
    private static final Logger LOG =
        LoggerFactory.getLogger(PLAH.class);

    private Map<String, List<Event>> logIdToEvents;
    private int totalLogFiles;
    private Set<String> eventNames = new
        TreeSet<String>();
    private Set<Pair<String, String>>
        eventDependency =                    new
        HashSet<Pair<String, String>>();
    private Map<String, Integer>
        eventNameToFrequency =
        new HashMap<String, Integer>();

    // And correlation
    private Map<Pair<String, String>, Boolean>
        andCorrelation;
    private Map<Pair<String, String>, Integer>
        andCorrelationCount                  = new
        HashMap<Pair<String, String>, Integer>();
```

```java
private Set<String> andEvents = new
   HashSet<String>();

// Or correlation
private Set<Pair<String, String>> orCorrelation;
//Xor correlation
private Set<Pair<String, String>> xorCorrelation;
//parallel
private Set<Pair<String, String>>
   partialParallelCorrelation;
private Set<Pair<String, String>>
   fullParallelCorrelation;

//probability

public PLAH(Map<String, List<Event>>
   logIdToEvents) {
   this.logIdToEvents = logIdToEvents;
   this.totalLogFiles = logIdToEvents.size();
}

// Sort the events according to time stamp
private void sortEvents(List<Event> events) {
   Collections.sort(events, new
      Event.EventTsComparator());
}

private void probabilityLearning() {
   LOG.info("=============================");
   LOG.info("events probability");
   for (Map.Entry<String, Integer> entry :
      eventNameToFrequency.entrySet()) {
      LOG.info("{}  = {}", entry.getKey(), 1.0
         * entry.getValue() / totalLogFiles);
   }
   LOG.info("=============================");

   LOG.info("=============================");
   LOG.info("and correlation probability");
   for (Map.Entry<Pair<String, String>,
      Boolean> entry :
      andCorrelation.entrySet()) {
      if (entry.getValue()) {
         LOG.info("{}  = {}", entry.getKey(),
            1.0 *
            andCorrelationCount.get(entry.getKey())
            / totalLogFiles);
      }
```

```
        }
        LOG.info("============================");
}


public void run() {
    for (String logId : logIdToEvents.keySet()) {
        List<Event> events =
            logIdToEvents.get(logId);
        sortEvents(events);
        eventAndDependancyLearing(events);
        andCorrelationLearing(events);
        LOG.debug("event names after processing
            log file {}: {}", logId, eventNames);
        LOG.debug("event relations after
            processing log file {}: {}", logId,
            eventDependency);
        LOG.debug("event names frequency after
            processing log file {}: {}", logId,
            eventNameToFrequency);
        LOG.debug("and correlation after
            processing log file {}: {}", logId,
            andCorrelation);
    }

    setTotalOrCorrelation();
    setTotalXorCorrelation();
    setTotalParallelCorrelation();
    for (String logId : logIdToEvents.keySet()) {
        List<Event> events =
            logIdToEvents.get(logId);
        orCorrelationLearing(events);
        xorCorrelationLearing(events);
        parallelLearning(events);
        LOG.debug("or correlation after
            processing log file {}:
        {}", logId, orCorrelation);
        LOG.debug("xor correlation after
            processing log file {}: {}", logId,
            xorCorrelation);
        LOG.debug("partial parallel correlation
            after processing log file {}: {}",
            logId, partialParallelCorrelation);
        LOG.debug("full parallel correlation
            after processing log file {}: {}",
            logId, fullParallelCorrelation);
    }
```

```
            probabilityLearning();
            printLearning();
    }

}
```

Another Class in the platform that connects any csv file (data file) with the algorithms steps:

```
  private CellProcessor[] getCellProcessors() {
        CellProcessor[] processors = new
            CellProcessor[]{
                new NotNull(), // logId
                new NotNull(), // eventName
                new ParseLong(), // start ts
                new ParseLong() // end ts
        };
        return processors;
    }
```

The class that run all the steps to give output result is named Driver:

```
package com.research.mining;

import org.slf4j.Logger;
import org.slf4j.LoggerFactory;

import java.util.List;
import java.util.Map;

public class Driver {
    private static final Logger LOG =
        LoggerFactory.getLogger(Driver.class);

    public static void main(String[] args) {
        if (args.length != 1) {
            LOG.error("command filename");
            System.exit(1);
        }
        CSVToEvents csvToEvents = new
            CSVToEvents(args[0]);
        Map<String, List<Event>> logIdToEvents =
            csvToEvents.readEvents();
        PLAH plah = new PLAH(logIdToEvents);
        plah.run();
    }
}
```

Zayoud et al. (2019a)

E MYL Software Output Examples

The output of the previous platform:

17:30:11.255 [main] DEBUG com.research.mining.PLAH - partial parallel correlation after processing log file 4: []

17:30:11.255 [main] DEBUG com.research.mining.PLAH - full parallel correlation after processing log file 4: []

17:30:11.256 [main] INFO com.research.mining.PLAH - ===============

17:30:11.256 [main] INFO com.research.mining.PLAH - events probability

17:30:11.258 [main] INFO com.research.mining.PLAH - er unit = 0.25

17:30:11.258 [main] INFO com.research.mining.PLAH - arrival = 0.25

17:30:11.258 [main] INFO com.research.mining.PLAH - release resources = 0.5

17:30:11.259 [main] INFO com.research.mining.PLAH - patient preparation = 0.25

17:30:11.259 [main] INFO com.research.mining.PLAH - discharge patient = 0.5

17:30:11.259 [main] INFO com.research.mining.PLAH - data collection = 0.25

17:30:11.259 [main] INFO com.research.mining.PLAH - treatment process = 0.5

17:30:11.259 [main] INFO com.research.mining.PLAH - billing = 0.25

17:30:11.259 [main] INFO com.research.mining.PLAH - ===============

17:30:11.259 [main] INFO com.research.mining.PLAH - ===============

17:30:11.259 [main] INFO com.research.mining.PLAH - and correlation probability

17:30:11.259 [main] INFO com.research.mining.PLAH - arrival=er unit = 0.25

17:30:11.259 [main] INFO com.research.mining.PLAH - data collection=er unit = 0.25

17:30:11.259 [main] INFO com.research.mining.PLAH - discharge patient=release resources = 0.5

17:30:11.259 [main] INFO com.research.mining.PLAH - arrival=data collection = 0.25

17:30:11.259 [main] INFO com.research.mining.PLAH - ===============

17:30:11.259 [main] INFO com.research.mining.PLAH - event names: [arrival, billing, data collection, discharge patient, er unit, patient preparation, release resources, treatment process]

17:30:11.259 [main] INFO com.research.mining.PLAH - event relations: [arrival=er unit, discharge patient=release resources, data collection=arrival]

17:30:11.260 [main] INFO com.research.mining.PLAH - event names frequency: er unit=1, arrival=1, release resources=2, patient preparation=1, discharge patient=2, data collection=1, treatment process=2, billing=1

DOI: 10.1201/9781003366577-E

17:30:11.260 [main] INFO com.research.mining.PLAH - and correlation: arrival=release resources=false, arrival=er unit=true, data collection=er unit=true, discharge patient=er unit=false, discharge patient=release resources=true, data collection=discharge patient=false, billing=treatment process=false, er unit=release resources=false, arrival=data collection=true, data collection=release resources=false, patient preparation=treatment process=false, arrival=discharge patient=false

17:30:11.260 [main] INFO com.research.mining.PLAH - or correlation: [discharge patient=treatment process, release resources=treatment process]

References

Abraham, A. and Das, S. (2010). *Computational intelligence in power engineering*, volume 302. Springer.

Agaoglu, M. (2016). Predicting instructor performance using data mining techniques in higher education. *IEEE Access*, 4:2379–2387.

Agrawal, R., Gunopulos, D., and Leymann, F. (1998). Mining process models from workflow logs. In *International Conference on Extending Database Technology*, pages 467–483. Springer.

Agrawal, R., Srikant, R., et al. (1995). Mining sequential patterns. In *icde*, volume 95, pages 3–14.

Aha, D. W., Kibler, D., and Albert, M. K. (1991a). Instance-based learning algorithms. *Machine learning*, 6(1):37–66.

Aha, D. W., Kibler, D., and Albert, M. K. (1991b). Instance-based learning algorithms. *Machine learning*, 6(1):37–66.

Akerkar, R. (2014). Big data computing. florida, usa: Crc press. In *Al Nuaimi, E., Al Neyadi, H., Mohamed, N., & Al-Jaroodi, J. (2015). Applications of big data to smart cities. Journal of Internet Services and Applications, 6(1), 1–15*. Taylor & Francis Group.

Akshay, V. (2012). Object-oriented framework for healthcare simulation.

Alves de Medeiros, A., Van Dongen, B., Van Der Aalst, W., and Weijters, A. (2004a). Process mining: Extending the alpha-algorithm to mine short loops. *University of Technology, Eindhoven*, 113:145–180.

Alves de Medeiros, A., Van Dongen, B., Van Der Aalst, W., and Weijters, A. (2004b). Process mining: Extending the alpha-algorithm to mine short loops. *University of Technology, Eindhoven*, 113:145–180.

Alves de Medeiros, A. K. (2006). Genetic process mining. *CIP-DATA LIBRARY TECHNISCHE UNIVERSITEIT EINDHOVEN, printed in*.

Anbarasi, M., Anupriya, E., and Iyengar, N. (2010). Enhanced prediction of heart disease with feature subset selection using genetic algorithm. *International Journal of Engineering Science and Technology*, 2(10):5370–5376.

Antão, A. and Grenha, C. (2012). Estudo da viabilidade das farmacias. *Edited by Oliveira Reis &*, 469.

Aveiro, D., Silva, A. R., and Tribolet, J. (2010). Towards a god-theory for organizational engineering: continuously modeling the continuous (re) generation, operation and deletion of the enterprise. In *Proceedings of the 2010 ACM Symposium on Applied Computing*, pages 150–157.

Banerjee, A. and Gupta, P. (2015). Extension to alpha algorithm for process mining. *Ijecs. in.*, 4(9):14383–14386.

Bartsch-Spörl, B., Lenz, M., and Hübner, A. (1999). Case-based reasoning–survey and future directions. In *German Conference on Knowledge-Based Systems*, pages 67–89. Springer.

Bates, D. W., Saria, S., Ohno-Machado, L., Shah, A., and Escobar, G. (2014). Big data in health care: using analytics to identify and manage high-risk and high-cost patients. *Health Affairs*, 33(7):1123–1131.

Breiman, L. (1984). Classification and regression trees. Technical report.

Breiman, L. and Stone, C. (1978). Parsimonious binary classification trees. *Technology Service Corporation, Santa Monica, Calif. Tech. Rep. TSCCSD-TN*, 4.

Brin, S. and Page, L. (1998). The anatomy of a large-scale hypertextual web search engine. *Computer Networks and ISDN Systems*, 30(1-7):107–117.

Burattin, A. (2015). Process mining techniques in business environments. In *volume 207 of Lecture Notes in Business Information Processing*. Springer.

Burattin, A., Sperduti, A., and Veluscek, M. (2013). Business models enhancement through discovery of roles. In *2013 IEEE Symposium on Computational Intelligence and Data Mining (CIDM)*, pages 103–110. IEEE.

Caetano, A., Assis, A., and Tribolet, J. (2011). Using business transactions to analyse the consistency of business process models. In *2012 45th Hawaii International Conference on System Sciences*, pages 4277–4285. IEEE.

Chapelle, O., Haffner, P., and Vapnik, V. N. (1999). Support vector machines for histogram-based image classification. *IEEE Transactions on Neural Networks*, 10(5):1055–1064.

Chen, Y.-Y., Gan, Q., and Suel, T. (2002). I/o-efficient techniques for computing pagerank. In *Proceedings of the eleventh international conference on Information and knowledge management*, pages 549–557.

Christensen, C. M., Grossman, J. H., and Hwang, J. (2009). *The innovator's prescription: a disruptive solution for health care*.

Chui, C.-K., Kao, B., and Hung, E. (2007). Mining frequent itemsets from uncertain data. In *Pacific-Asia Conference on Knowledge Discovery and Data Mining*, pages 47–58. Springer.

Chui, M., Brown, B., Bughin, J., Dobbs, R., Roxburgh, C., and Manyika, A. (2011). Big Data: The Next Frontier for Innovation, Competition, and Productivity. *McKinsey Global Institute*.

Chung, W. (2014). Bizpro: Extracting and categorizing business intelligence factors from textual news articles. *International Journal of Information Management*, 34(2):272–284.

Comission of the European Communities (2007). Remuneration of researchers in the public and private sectors.

Conforti, R., La Rosa, M., and ter Hofstede, A. H. (2017). Filtering out infrequent behavior from business process event logs. *IEEE Transactions on Knowledge and Data Engineering*, 29(2):300–314.

Cook, J. E. and Wolf, A. L. (1998). Discovering models of software processes from event-based data. *ACM Transactions on Software Engineering and Methodology (TOSEM)*, 7(3):215–249.

Costa, A. P., Linhares, R., and Neri de Souza, F. (2012). Possibilidades de análise qualitativa no webqda e colaboração entre pesquisadores em educação em comunicação. *Anais 3° Simpósio Educação e Comunicação: Infoinclusão possibilidades de ensinar e aprender*, pages 276–286.

Cristianini, N., Campbell, C., and Shawe-Taylor, J. (1999). A multiplicative updating algorithm for training support vector machine. In *In 6th European Symposium on Arti⁻ cial Neural Networks (ESANN*. Citeseer.

Data (2010). Data everywhere.

Datta, A. (1998). Automating the discovery of as-is business process models: probabilistic and algorithmic approaches. *Information Systems Research*, 9(3):275–301.

De Medeiros, A. A. and Weijters, A. (2005). Genetic process mining. In *Applications and Theory of Petri Nets 2005, Volume 3536 of Lecture Notes in Computer Science*. Citeseer.

Dehnert, J. (2003). A methodology for workflow modeling. *From business process modeling towards sound workflow specification: Dissertation/Technische Universitat Berlin, Germany.*

Dempster, A. P., Laird, N. M., and Rubin, D. B. (1977). Maximum likelihood from incomplete data via the em algorithm. *Journal of the Royal Statistical Society: Series B (Methodological)*, 39(1):1–38.

der Aalst, V. (2011a). Process discovery: an introduction. In *Process Mining*, pages 125–156. Springer.

der Aalst, V. (2011b). Process discovery: an introduction. In *Process Mining*, pages 125–156. Springer.

Devillers, R. and Antti, V. (2015). *Application and theory of petri nets and concurrency*. Springer.

Dittrich, J. and Quiané-Ruiz, J.-A. (2012). Efficient big data processing in hadoop mapreduce. *Proceedings of the VLDB Endowment*, 5(12):2014–2015.

Domingos, P. and Pazzani, M. (1997). On the optimality of the simple bayesian classifier under zero-one loss. *Machine learning*, 29(2-3):103–130.

dos Santos Garcia, C., Meincheim, A., Junior, E. R. F., Dallagassa, M. R., Sato, D. M. V., Carvalho, D. R., Santos, E. A. P., and Scalabrin, E. E. (2019). Process mining techniques and applications-a systematic mapping study. *Expert Systems with Applications*.

Drucker, H., Wu, D., and Vapnik, V. N. (1999). Support vector machines for spam categorization. *IEEE Transactions on Neural networks*, 10(5):1048–1054.

Duda, R. O., Hart, P. E., and Stork, D. G. (2012). *Pattern classification*. John Wiley & Sons.

Duda, R. O., Hart, P. E., Stork, D. G., et al. (2001). Pattern classification. *International Journal of Computational Intelligence and Applications*, 1:335–339.

Dumais, S., Platt, J., Heckerman, D., and Sahami, M. (1998). Inductive learning algorithms and representations for text categorization. In *Proceedings of the seventh international conference on Information and knowledge management*, pages 148–155. ACM.

Dumas, M., van der Aalst, W., and Ter Hofstede, A. (2005). *Process aware information systems*, volume 1. Wiley Online Library.

Durrett, R. (2010a). *Probability: Theory and Examples*. Cambridge University Press.

Durrett, R. (2010b). *Probability: Theory and Examples*. Cambridge University Press.

D'Alessandro, M., Gaudiano, G., Viola, G., and Santarcangelo, V. (1999). How to extract workflows from data: process mining.

Evermann, J. and Assadipour, G. (2014). Big data meets process mining: implementing the alpha algorithm with map-reduce. In *Proceedings of the 29th Annual ACM Symposium on Applied Computing*, pages 1414–1416. ACM.

Fan, J., Han, F., and Liu, H. (2014). Challenges of big data analysis. *National Science Review*, 1(2):293–314.

Fan, J. and Lv, J. (2008). Sure independence screening for ultrahigh dimensional feature space. *Journal of the Royal Statistical Society: Series B (Statistical Methodology)*, 70(5):849–911.

Fayyad, U. M., Piatetsky-Shapiro, G., Smyth, P., Uthurusamy, R., et al. (1996). *Advances in Knowledge Discovery and Data Mining*, volume 21. AAAI Press Menlo Park.

Fix, E. and Hodges Jr, J. L. (1951). Discriminatory analysis-nonparametric discrimination: small sample performance. Technical report, California Univ Berkeley.

Fix, E. and Hodges Jr, J. L. (1952). Discriminatory analysis-nonparametric discrimination: Small sample performance. Technical report, California Univ Berkeley.

Freedman, L. P. (2005). Achieving the mdgs: health systems as core social institutions. *Development*, 48(1):19–24.

Friedman, J. H. (1977a). A recursive partitioning decision rule for nonparametric classification. *IEEE Transactions on Computers*, (4):404–408.

Friedman, J. H. (1977b). A recursive partitioning decision rule for nonparametric classification. *IEEE Transactions on Computers*, (4):404–408.

Friedman, N., Geiger, D., and Goldszmidt, M. (1997). Bayesian network classifiers. *Machine Learning*, 29(2-3):131–163.

Frith, U. and Happé, F. (2005). Autism spectrum disorder. *Current Biology*, 15(19):R786–R790.

Fung, G. and Stoeckel, J. (2007). Svm feature selection for classification of spect images of alzheimer's disease using spatial information. *Knowledge and Information Systems*, 11(2):243–258.

Gandomi, A. and Haider, M. (2015a). Beyond the hype: Big data concepts, methods, and analytics. *International Journal of Information Management*, 35(2):137–144.

Gandomi, A. and Haider, M. (2015b). Beyond the hype: Big data concepts, methods, and analytics. *International Journal of Information Management*, 35(2):137–144.

Ge, Z., Song, Z., Ding, S. X., and Huang, B. (2017). Data mining and analytics in the process industry: the role of machine learning. *IEEE Access*, 5:20590–20616.

Géczy, P. (2014). Big data characteristics. *The Macrotheme Review*, 3(6):94–104.

Ghosh, J. and Liu, A. (2009). K-means. In *The Top Ten Algorithms in Data Mining*, pages 35–50. Chapman and Hall/CRC.

Ghosh, J. and Strehl, A. (2006). Similarity-based text clustering: a comparative study. In *Grouping Multidimensional Data*, pages 73–97. Springer.

Gibson, U., Heid, C. A., and Williams, P. M. (1996). A novel method for real time quantitative rt-pcr. *Genome Research*, 6(10):995–1001.

Gilson, L. (2003). Trust and the development of health care as a social institution. *Social Science & Medicine*, 56(7):1453–1468.

Gonella, P. (2016). Process mining: a database of applications. Technical report.

Gray, R. M. and Neuhoff, D. L. (1998). Quantization. *IEEE Transactions on Information Theory*, 44(6):2325–2383.

Grol, R. and Grimshaw, J. (1999). Evidence-based implementation of evidence-based medicine. *The Joint Commission Journal on Quality Improvement*, 25(10):503–513.

Gundecha, P. and Liu, H. (2012). Mining social media: a brief introduction. tutorials in operations research. *Inform*, 1(4).

Günther, C. W. and Van Der Aalst, W. M. (2007a). Fuzzy mining–adaptive process simplification based on multi-perspective metrics. In *International Conference on Business Process Management*, pages 328–343. Springer.

Günther, C. W. and Van Der Aalst, W. M. (2007b). Fuzzy mining–adaptive process simplification based on multi-perspective metrics. In *International Conference on Business Process Management*, pages 328–343. Springer.

Günther, C. W. and Van Der Aalst, W. M. (2007c). Fuzzy mining–adaptive process simplification based on multi-perspective metrics. In *International Conference on Business Process Management*, pages 328–343. Springer.

Günther, C. W. and Van Der Aalst, W. M. (2007d). Fuzzy mining–adaptive process simplification based on multi-perspective metrics. In *International Conference on Business Process Management*, pages 328–343. Springer.

Hakeem, A., Gupta, H., Kanaujia, A., Choe, T. E., Gunda, K., Scanlon, A., Yu, L., Zhang, Z., Venetianer, P., Rasheed, Z., et al. (2012). Video analytics for business intelligence. In *Video Analytics for Business Intelligence*, pages 309–354. Springer.

Hand, D. J. and Yu, K. (2001). Idiot's bayes—not so stupid after all? *International Statistical Review*, 69(3):385–398.

Health system (2017). Health system — Wikipedia, the free encyclopedia. [Online; accessed 01-September-2017].

Henriques, M., Tribolet, J., and Hoogervorst, J. (2010). Enterprise governance and demo: a reference method to guide enterprise (re) design and operation with demo. In *Conferência da Associação Portuguesa de Sistemas de Informação, Viana do Castelo*, volume 1, pages 1–13.

Hirschberg, J., Hjalmarsson, A., and Elhadad, N. (2010). "you're as sick as you sound": Using computational approaches for modeling speaker state to gauge illness and recovery. In *Advances in Speech Recognition*, pages 305–322. Springer.

Holmes, A. (2012). *Hadoop in Practice*. Manning Publications Co.

Hoogervorst, J. A. and Dietz, J. L. (2008). Enterprise architecture in enterprise engineering. *Enterprise Modelling and Information Systems Architectures (EMISAJ)*, 3(1):3–13.

Hornỳ, M. (2014). Bayesian networks. Technical report, Technical report.

Hossain, M. S. and Muhammad, G. (2016). Healthcare big data voice pathology assessment framework. *iEEE Access*, 4:7806–7815.

Huang, K., Yang, H., King, I., and Lyu, M. R. (2004). Learning large margin classifiers locally and globally. In *Proceedings of the Twenty-First International Conference on Machine Learning*. ICML.

Hunt, E. B., Marin, J., and Stone, P. J. (1966). Experiments in induction.

Jamain, A. and Hand, D. J. (2008). Mining supervised classification performance studies: A meta-analytic investigation. *Journal of Classification*, 25(1):87–112.

Jans, M., Van Der Werf, J. M., Lybaert, N., and Vanhoof, K. (2011). A business process mining application for internal transaction fraud mitigation. *Expert Systems with Applications*, 38(10):13351–13359.

Jansen-Vullers, M. and Reijers, H. (2005). Business process redesign in healthcare: towards a structured approach. *INFOR: Information Systems and Operational Research*, 43(4):321–339.

Jensen, A., Sharif, H., Svare, E. I., Frederiksen, K., and Kjær, S. K. (2007). Risk of breast cancer after exposure to fertility drugs: results from a large danish cohort study. *Cancer Epidemiology and Prevention Biomarkers*, 16(7):1400–1407.

Jia, Y., Zhang, J., and Huan, J. (2011). An efficient graph-mining method for complicated and noisy data with real-world applications. *Knowledge and Information Systems*, 28(2):423–447.

Jiang, J. (2012). Information extraction from text. In *Mining Text Data*, pages 11–41. Springer.

Kaplan, R. S. and Porter, M. E. (2011). How to solve the cost crisis in health care. *Harvard Business Review*, 89(9):46–52.

Kassirer, J. P., Kopelman, R. I., and Wong, J. B. (1991). *Learning Clinical Reasoning*. Williams and Wilkins Baltimore.

Katal, A., Wazid, M., and Goudar, R. (2013). Big data: issues, challenges, tools and good practices. In *Contemporary Computing (IC3), 2013 Sixth International Conference on*, pages 404–409. IEEE.

Kaymak, U., Mans, R., van de Steeg, T., and Dierks, M. (2012). On process mining in health care. In *Systems, Man, and Cybernetics (SMC), 2012 IEEE International Conference on*, pages 1859–1864. IEEE.

Kearns, M. J. (1999). *Advances in Neural Information Processing Systems 11: Proceedings of the 1998 Conference*, volume 11. MIT Press.

Keller, G. and Teufel, T. (1998). *SAP R/3 Process Oriented Implementation*. Addison-Wesley Longman Publishing Co., Inc.

Kononenko, I. (1993). Inductive and bayesian learning in medical diagnosis. *Applied Artificial Intelligence an International Journal*, 7(4):317–337.

Koski, T. and Noble, J. (2011). *Bayesian Networks: An Introduction*, volume 924. John Wiley & Sons.

Labrinidis, A. and Jagadish, H. V. (2012). Challenges and opportunities with big data. *Proceedings of the VLDB Endowment*, 5(12):2032–2033.

Lapão, L. V. and Dussault, G. (2012). From policy to reality: clinical managers' views of the organizational challenges of primary care reform in portugal. *The International Journal of Health Planning and Management*, 27(4):295–307.

Laudon, K. C. and Laudon, J. P. (1999). *Management Information Systems*. Prentice Hall PTR.

Lee, J., Bagheri, B., and Kao, H.-A. (2014). Recent advances and trends of cyber-physical systems and big data analytics in industrial informatics. In *International Proceeding of International Conference on Industrial Informatics (INDIN)*, pages 1–6.

Lerman, I.-C. (2000). Comparing taxonomic data. in proceeding of ecml-2000 workshop dealing with structured data. *Mathématiques et Sciences Humaines. Mathematics and Social Sciences*, (151):18–29.

Leung, E. W., Medeiros, F. A., and Weinreb, R. N. (2008). Prevalence of ocular surface disease in glaucoma patients. *Journal of Glaucoma*, 17(5):350–355.

Lewis, D. D., Yang, Y., Rose, T. G., and Li, F. (2004). Rcv1: A new benchmark collection for text categorization research. *Journal of Machine Learning Research*, 5(Apr):361–397.

Lifvergren, S., Docherty, P., and Hellström, A. (2010). Management by dialogue: joint reflection, sense making and development. In *Cornell University's International Health Care Conference–A Time for Change: Restructuring America's Health Care Delivery System, New York, NY, May*, pages 11–12.

Lloyd, S. (1982). Least squares quantization in pcm. *IEEE Transactions on Information Theory*, 28(2):129–137.

MacQueen, J. et al. (1967a). Some methods for classification and analysis of multivariate observations. In *Proceedings of the Fifth Berkeley Symposium on Mathematical Statistics and Probability*, volume 1, pages 281–297. Oakland, CA, USA.

MacQueen, J. et al. (1967b). Some methods for classification and analysis of multivariate observations. In *Proceedings of the Fifth Berkeley Symposium on Mathematical Statistics and Probability*, volume 1, pages 281–297. Oakland, CA, USA.

Mani, S., Pazzani, M. J., and West, J. (1997). Knowledge discovery from a breast cancer database. In *Conference on Artificial Intelligence in Medicine in Europe*, pages 130–133. Springer.

Mans, R. S., van der Aalst, W. M. P., and Vanwersch, R. J. B. (2015a). Process mining in healthcare: Evaluating and exploiting operational healthcare processes.

Mans, R. S., Schonenberg, M., Song, M., van der Aalst, W. M. P., and Bakker, P. J. (2008a). Application of process mining in healthcare–a case study in a dutch hospital. In *International Joint Conference on Biomedical Engineering Systems and Technologies*, pages 425–438. Springer.

Mans, R. S., Schonenberg, M., Song, M., van der Aalst, W. M. P., and Bakker, P. J. (2008b). Application of process mining in healthcare–a case study in a dutch hospital. In *International Joint Conference on Biomedical Engineering Systems and Technologies*, pages 425–438. Springer.

Mans, R. S., Van der Aalst, W. M., and Vanwersch, R. J. (2015b). *Process Mining in Healthcare: Evaluating and Exploiting Operational Healthcare Processes.* Springer.

Mayer, R. E. (1996). Learners as information processors: Legacies and limitations of educational psychology's second.. *Educational Psychologist*, 31(3-4):151–161.

MIKE2.0 (2013). Big data definition.

Mintzberg, H. et al. (1994). The fall and rise of strategic planning. *Harvard Business Review*, 72(1):107–114.

Moreno, J. L. (1934). Who shall survive? A new approach to the problem of human interrelations.

Murata, T. (1989a). Petri nets: properties, analysis and applications. *Proceedings of the IEEE*, 77(4):541–580.

Murata, T. (1989b). Petri nets: properties, analysis and applications. *Proceedings of the IEEE*, 77(4):541–580.

Murphy, K. (1998). A brief introduction to graphical models and bayesian networks.

Nemati, H. R. and Barko, C. D. (2004). *Organizational Data Mining: Leveraging Enterprise Data Resources for Optimal Performance.* IGI Global.

O'Brien, J. A., Marakas, G. M., et al. (2011). *Management Information Systems*, volume 9. McGraw-Hill/Irwin.

OECD (2012). *OECD Public Governance Reviews Slovenia: Towards a Strategic and Efficient State.* Organization for Economic.

OMG (2011). Notation (bpmn). *FTF Beta*, 1.

Oueida, S., Kadry, S., and Abichar, P. (2017). Emergency department proposed model: Estimating kpis. In *International Conference on Management and Industrial Engineering*, number 8, pages 390–403. Niculescu Publishing House.

Oueida, S., Kadry, S., Abichar, P., and Ionescu, S. (2018). The applications of simulation modeling in emergency departments: a review. In *Health Care Delivery and Clinical Science: Concepts, Methodologies, Tools, and Applications*, pages 1014–1045. IGI Global.

Page, L., Brin, S., Motwani, R., and Winograd, T. (1999). The pagerank citation ranking: Bringing order to the web. technical report 1999–0120. Technical report, Computer Science Department, Stanford University.

Papadimitriou, S. and Sun, J. (2008). Disco: Distributed co-clustering with map-reduce: A case study towards petabyte-scale end-to-end mining. In *Data Mining, 2008. ICDM'08. Eighth IEEE International Conference on*, pages 512–521. IEEE.

Patil, H. A. (2010). "cry baby": Using spectrographic analysis to assess neonatal health status from an infant's cry. In *Advances in Speech Recognition*, pages 323–348. Springer.

Petri, C. A. (1962). *Kommunikationen mit automaten.* PhD thesis, PhD Thesis, University of Bonn, 1962. English translation: Technical Report . . . .

Petri, C. A. and Reisig, W. (2008). Petri net. *Scholarpedia*, 3(4):6477.

Process mining (2017). Process mining — Wikipedia, the free encyclopedia. [On-line; accessed 15-August-2006].

Public Health Evaluation and Impact Assessment Consortium (2011). Specific contract: Mid-term evaluation of the health programme.

Quinlan, J. R. (1986). Induction of decision trees. *Machine Learning*, 1(1):81–106.

Quinlan, J. R. (1993). C4.5: Programs for machine learning.

Radnor, Z. J., Holweg, M., and Waring, J. (2012). Lean in healthcare: the unfilled promise? *Social Science & Medicine*, 74(3):364–371.

Ramakrishnan, N. (2009a). C4. 5. In *The Top Ten Algorithms in Data Mining*, pages 15–34. Chapman and Hall/CRC.

Ramakrishnan, S. (2009b). Genetic algorithm based approach to access structure selection with storage constraint. US Patent 7,620,609.

Rebuge, Á. and Ferreira, D. R. (2012). Business process analysis in healthcare environments: A methodology based on process mining. *Information Systems*, 37(2):99–116.

Reisig, W. and Rozenberg, G. (1998). *Lectures on petri nets i: basic models: advances in petri nets*. Springer Science & Business Media.

Rojas, E., Munoz-Gama, J., Sepúlveda, M., and Capurro, D. (2016a). Process mining in healthcare: A literature review. *Journal of Biomedical Informatics*, 61:224–236.

Rojas, E., Munoz-Gama, J., Sepúlveda, M., and Capurro, D. (2016b). Process mining in healthcare: A literature review. *Journal of Biomedical Informatics*, 61:224–236.

Rovani, M., Maggi, F. M., de Leoni, M., and van der Aalst, W. M. (2015). Declarative process mining in healthcare. *Expert Systems with Applications*, 42(23):9236–9251.

Rubin, V., Günther, C. W., Van Der Aalst, W. M., Kindler, E., Van Dongen, B. F., and Schäfer, W. (2007). Process mining framework for software processes. In *International Conference on Software Process*, pages 169–181. Springer.

Rumbaugh, Jamesand Booch, G. and Jacobson, I. (1999). *The Unified Software Development*. Process.

Sabbaghi, A. and Vaidyanathan, G. (2004). Swot analysis and theory of constraint in information technology projects. *Information systems Education Journal*, 2(23):3–19.

Sani, M. F., van Zelst, S. J., and van der Aalst, W. M. (2018). Applying sequence mining for outlier detection in process mining. In *OTM Confederated International Conferences" On the Move to Meaningful Internet Systems"*, pages 98–116. Springer.

Scholkopf, B. (1997). *Support Vector Learning*. R. Oldenbourg Verlag.

Schölkopf, B., Burges, C. J., and Smola, A. J. (1999a). Introduction to support vector learning. In *Advances in Kernel Methods*, pages 1–15. MIT Press.

Schölkopf, B., Burges, C. J., Smola, A. J., et al. (1999b). *Advances in kernel methods: support vector learning*.

Scott, J. (2012). Social network analysis. *Lectures, Comptes rendus à paraître*.

Sivarajah, U., Kamal, M. M., Irani, Z., and Weerakkody, V. (2017). Critical analysis of big data challenges and analytical methods. *Journal of Business Research*, 70:263–286.

Smola, A. J., Bartlett, P., Schölkopf, B., and Schuurmans, D. (1999). Advances in large margin classifiers.

Smola, A. J. and Schölkopf, B. (2004). A tutorial on support vector regression. *Statistics and Computing*, 14(3):199–222.

Soni, J., Ansari, U., Sharma, D., and Soni, S. (2011). Predictive data mining for medical diagnosis: An overview of heart disease prediction. *International Journal of Computer Applications*, 17(8):43–48.

Starfield, B. (1998). Primary care visits and health policy. *CMAJ: Canadian Medical Association Journal*, 159(7):795.

Steinbach, M. and Tan, P.-N. (2009). knn: k-nearest neighbors. In *The Top Ten Algorithms in Data Mining*, pages 165–176. Chapman and Hall/CRC.

Sun, W.-J., Pan, M.-Y., and Ye, Q. (2011). Improved association rule mining method based on t statistical. *Jisuanji Yingyong Yanjiu*, 28(6):2073–2077.

Szatmari, P. (2003). The causes of autism spectrum disorders: Multiple factors have been identified, but a unifying cascade of events is still elusive. *BMJ*, 326:173–174.

Tan, P.-N., Steinbach, M., and Kumar, V. (2006). Classification: basic concepts, decision trees, and model evaluation. *Introduction to Data Mining*, 1:145–205.

The Open Group (2009). *Soa source book*. Van Haren Publishing.

Titterington, D., Murray, G., Murray, L., Spiegelhalter, D., Skene, A., Habbema, J., and Gelpke, G. (1981). Comparison of discrimination techniques applied to a complex data set of head injured patients. *Journal of the Royal Statistical Society: Series A (General)*, 144(2):145–161.

Tole, A. A. et al. (2013). Big data challenges. *Database Systems Journal*, 4(3):31–40.

Tribunal de Contas (2011). Normas de auditoria do tribunal de contas da união. revisão junho 2011. *Boletim do Tribunal de Contas da União*.

Turhan, B. and Bener, A. (2009). Analysis of naive bayes' assumptions on software fault data: an empirical study. *Data & Knowledge Engineering*, 68(2):278–290.

van Aalst, W. M., van Hee, K. M., van Werf, J. M., and Verdonk, M. (2010). Auditing 2.0: Using process mining to support tomorrow's auditor. *Computer*, 43(3).

van der Aalst, W. (2009). Process mining: beyond business intelligence.

Van Der Aalst, W. (2011). *Process mining: discovery, conformance and enhancement of business processes*, volume 2. Springer.

Van Der Aalst, W. (2012). Process mining: overview and opportunities. *ACM Transactions on Management Information Systems (TMIS)*, 3(2):1–17.

Van Der Aalst, W., Adriansyah, A., De Medeiros, A. K. A., Arcieri, F., Baier, T., Blickle, T., Bose, J. C., Van Den Brand, P., Brandtjen, R., Buijs, J., et al. (2011). Process mining manifesto. In *International Conference on Business Process Management*, pages 169–194. Springer.

Van der Aalst, W., Weijters, T., and Maruster, L. (2004a). Workflow mining: discovering process models from event logs. *IEEE Transactions on Knowledge & Data Engineering*, (9):1128–1142.

Van der Aalst, W., Weijters, T., and Maruster, L. (2004b). Workflow mining: discovering process models from event logs. *IEEE Transactions on Knowledge and Data Engineering*, 16(9):1128–1142.

Van der Aalst, W., Weijters, T., and Maruster, L. (2004c). Workflow mining: discovering process models from event logs. *IEEE Transactions on Knowledge & Data Engineering*, (9):1128–1142.

Van der Aalst, W., Weijters, T., and Maruster, L. (2004d). Workflow mining: discovering process models from event logs. *IEEE Transactions on Knowledge & Data Engineering*, (9):1128–1142.

Van der Aalst, W. M. (1998). The application of petri nets to workflow management. *Journal of Circuits, Systems, and Computers*, 8(01):21–66.

van Der Aalst, W. M. (2016). Process Mining - Data Science in Action. Berlin, Germany: Springer-Verlag Berlin Heidelberg.

van der Aalst, W. M., Bolt, A., and van Zelst, S. J. (2017). Rapidprom: mine your processes and not just your data. *arXiv preprint arXiv:1703.03740*.

van der Aalst, W. M., Rubin, V., van Dongen, B., Kindler, E., and Günther, C. (2009a). Process mining: a two-step approach using transition systems and regions. *Information Systems*, 34(3):305–327.

van der Aalst, W. M., Rubin, V., van Dongen, B. F., Kindler, E., and Günther, C. W. (2006a). Process mining: A two-step approach using transition systems and regions. *BPM Center Report BPM-06-30, BPMcenter.org*, 6.

van der Aalst, W. M., Rubin, V., van Dongen, B. F., Kindler, E., and Günther, C. W. (2006b). Process mining: A two-step approach using transition systems and regions. *BPM Center Report BPM-06-30, BPMcenter.org*, 6.

van der Aalst, W. M., Rubin, V., van Dongen, B. F., Kindler, E., and Günther, C. W. (2009b). Process mining: A two-step approach using transition systems and regions. *Information Systems*, 34:305–327.

Van der Aalst, W. M., Rubin, V., Verbeek, H., van Dongen, B. F., Kindler, E., and Günther, C. W. (2010). Process mining: a two-step approach to balance between underfitting and overfitting. *Software & Systems Modeling*, 9(1):87.

Van der Aalst, W. M. and van Dongen, B. F. (2002). Discovering workflow performance models from timed logs. In *International Conference on Engineering and Employment of Cooperative Information Systems*, pages 45–63. Springer.

Van der Aalst, W. M., van Dongen, B. F., Günther, C. W., Rozinat, A., Verbeek, E., and Weijters, T. (2009a). Prom: the process mining toolkit. *BPM (Demos)*, 489(31):2.

Van der Aalst, W. M., van Dongen, B. F., Günther, C. W., Rozinat, A., Verbeek, E., and Weijters, T. (2009b). Prom: the process mining toolkit. *BPM (Demos)*, 489(31):2.

Van der Aalst, W. M. and Weijters, A. (2004a). Process mining: a research agenda.

Van der Aalst, W. M. and Weijters, A. J. (2004b). Process mining: a research agenda. *Computers in Industry*, 53(4):231–244.

van Dongen, B. F., Busi, N., Pinna, G., and van der Aalst, W. (2007a). An iterative algorithm for applying the theory of regions in process mining. In *Proceedings of the Workshop on Formal Approaches to Business Processes and Web Services (FABPWS'07)*, pages 36–55. Publishing House of University of Podlasie, Siedlce, Poland.

van Dongen, B. F., Busi, N., Pinna, G., and van der Aalst, W. M. (2007b). An iterative algorithm for applying the theory of regions in process mining. In *Proceedings of the Workshop on Formal Approaches to Business Processes and Web Services (FABPWS07)*, pages 36–55. Publishing House of University of Podlasie, Siedlce, Poland.

Van Dongen, B. F. and Van der Aalst, W. M. (2004). Multi-phase process mining: Building instance graphs. In *International Conference on Conceptual Modeling*, pages 362–376. Springer.

van Dongen, B. F. and Van der Aalst, W. M. (2005). Multi-phase process mining: Aggregating instance graphs into epcs and petri nets. In *PNCWB 2005 workshop*, pages 35–58. Citeseer.

van Hee, K. M. (1994). *Information Systems Engineering: A Formal Approach.* Cambridge University Press.

Vapnik, V. and Vapnik, V. (1998). Statistical Learning Theory Wiley. *New York*, pages 156–160.

Vapnik, V. N. (1995). The nature of statistical learning theory.

Verbeek, H., Hirnschall, A., and van der Aalst, W. M. (2002). Xrl/flower: Supporting inter-organizational workflows using xml/petri-net technology. In *International Workshop on Web Services, E-Business, and the Semantic Web*, pages 93–108. Springer.

Walshe, D., Gartlan, J., Smith, A., Clennett, S., Tomlinson-Smith, A., Boas, L., and Robinson, A. (2010). An audit of the adequacy of acute wound care documentation of surgical inpatients. *Journal of Clinical Nursing*, 19(15–16):2207–2214.

Wang, J., Wong, R., Ding, J., Guo, Q., and Wen, L. (2013). Efficient selection of process mining algorithms. *IEEE Transactions on Services Computing*, 6(4).

Wang, J., Wong, R. K., Ding, J., Guo, Q., and Wen, L. (2012a). Efficient selection of process mining algorithms. *IEEE Transactions on Services Computing*, 6(4):484–496.

Wang, J., Wong, R. K., Ding, J., Guo, Q., and Wen, L. (2012b). Efficient selection of process mining algorithms. *IEEE Transactions on Services Computing*, 6(4):484–496.

Wasserman, S., Faust, K., et al. (1994). *Social Network Analysis: Methods and Applications*, volume 8. Cambridge University Press.

Weber, P., Bordbar, B., Tiňo, P., and Majeed, B. (2011). A framework for comparing process mining algorithms. In *2011 IEEE GCC Conference and Exhibition (GCC)*, pages 625–628. IEEE.

Weijters, A. and Ribeiro, J. (2011). Flexible heuristics miner (fhm). In *2011 IEEE Symposium on Computational Intelligence and Data Mining (CIDM)*, pages 310–317. IEEE.

Weijters, A., van Der Aalst, W. M., and De Medeiros, A. A. (2006a). Process mining with the heuristics miner-algorithm. *Technische Universiteit Eindhoven, Tech. Rep. WP*, 166:1–34.

Weijters, A., van Der Aalst, W. M., and De Medeiros, A. A. (2006b). Process mining with the heuristics miner-algorithm. *Technische Universiteit Eindhoven, Tech. Rep. WP*, 166:1–34.

Weijters, A., van Der Aalst, W. M., and De Medeiros, A. A. (2006c). Process mining with the heuristics miner-algorithm. *Technische Universiteit Eindhoven, Tech. Rep. WP*, 166:1–34.

Weijters, A., van Der Aalst, W. M., and De Medeiros, A. A. (2006d). Process mining with the heuristics miner-algorithm. *Technische Universiteit Eindhoven, Tech. Rep. WP*, 166:1–34.

Weijters, A., van Der Aalst, W. M., and De Medeiros, A. A. (2006e). Process mining with the heuristics miner-algorithm. *Technische Universiteit Eindhoven, Tech. Rep. WP*, 166:1–34.

Weijters, A., van Der Aalst, W. M., and De Medeiros, A. A. (2006f). Process mining with the heuristics miner-algorithm. *Technische Universiteit Eindhoven, Tech. Rep. WP*, 166:1–34.

Weijters, A. J. and Van der Aalst, W. M. (2003). Rediscovering workflow models from event-based data using little thumb. *Integrated Computer-Aided Engineering*, 10(2):151–162.

Won, H., Mah, W., and Kim, E. (2013). Autism spectrum disorder causes, mechanisms, and treatments: focus on neuronal synapses. *Frontiers in Molecular Neuroscience*, 6:19.

World Health Organization (2007). *Prevention of Cardiovascular Disease*. World Health Organization.

Wu, D., Liu, X., Hebert, S., Gentzsch, W., and Terpenny, J. (2015a). Performance evaluation of cloud-based high performance computing for finite element analysis. In *ASME 2015 International Design Engineering Technical Conferences and Computers and Information in Engineering Conference*, pages V01AT02A043–V01AT02A043. American Society of Mechanical Engineers.

Wu, D., Rosen, D. W., Wang, L., and Schaefer, D. (2015b). Cloud-based design and manufacturing: a new paradigm in digital manufacturing and design innovation. *Computer-Aided Design*, 59:1–14.

Wu, X. and Kumar, V. (2009). *The Top Ten Algorithms in Data Mining*. CRC Press.

Wu, X., Kumar, V., Ross Quinlan, J., Ghosh, J., Yang, Q., Motoda, H., McLachlan, G., Ng, A., Liu, B., Yu, P., et al. (2008). Top 10 algorithms in data mining. *Knowledge and Information Systems*, 14(1):1–37.

Xue, M. and Zhu, C. (2009). Applied research on data mining algorithm in network intrusion detection. In *2009 International Joint Conference on Artificial Intelligence*, pages 275–277. IEEE.

Yassine, A., Singh, S., and Alamri, A. (2017). Mining human activity patterns from smart home big data for health care applications. *IEEE Access*, 5:13131–13141.

Yeung, D. S., Wang, D., Ng, W. W., Tsang, E. C., and Wang, X. (2007a). Structured large margin machines: sensitive to data distributions. *Machine Learning*, 68(2):171–200.

Yeung, D. S., Wang, D., Ng, W. W., Tsang, E. C., and Wang, X. (2007b). Structured large margin machines: sensitive to data distributions. *Machine Learning*, 68(2):171–200.

Zayoud, M. and Ionescu, S. (2020). Improving healthcare industry using mining techniques. *Scientific Bulletin of Politehnica University of Bucharest*.

Zayoud, M., Kotb, Y., and Ionescu, S. (2018a). Algorithms for data and process mining. *FAIMA Business & Management Journal*, 6(2):45–56.

Zayoud, M., Kotb, Y., and Ionescu, S. (2019a). β algorithm: A new probabilistic process learning approach for big data in healthcare. *IEEE Access*, 99:78842–78869.

Zayoud, M., Kotb, Y., and Ionescu, S. (2019b). β algorithm: A new probabilistic process learning approach for big data in healthcare. *IEEE Access*, 7:78842–78869.

Zayoud, M., Kotb, Y., and Ionescu, S. (2019c). β algorithm: A new probabilistic process learning approach for big data in healthcare. *IEEE Access*, 7:78842–78869.

Zayoud, M., Oueida, S., and Kadry, S. (2018b). Wireless networks and security algorithms for efficient data encryption in smart healthcare. *Wireless Networks*.

Zayoud, M., Oueida, S., Kotb, Y., and AbiChar, P. (2017a). A probabilistic approach for threaded process learning. In *2017 IEEE 7th Annual Computing and Communication Workshop and Conference (CCWC)*, pages 1–6. IEEE.

Zayoud, M., Qotb, Y., and Ionescu, S. (2017b). Data and process mining techniques: survey literature. In *International Conference on Management and Industrial Engineering*, number 8, pages 145–157. Niculescu Publishing House.

Zayoud, M., Qotb, Y., and Ionescu, S. (2019d). Increasing economic benefits of organizations using appropriate process mining techniques. In *International Conference on Management and Industrial Engineering*, number 9. Niculescu Publishing House.

Zayoud, M., Qotb, Y., and Ionescu, S. (2019e). Using the β algorithm in improving health awareness. In *International Conference on Management and Industrial Engineering*, number 9. Niculescu Publishing House.

Zhang, J., Xue, N., and Huang, X. (2016). A secure system for pervasive social network-based healthcare. *IEEE Access*, 4:9239–9250.

Zhou, J., Li, X., and Shi, X. (2012). Long-term prediction model of rockburst in underground openings using heuristic algorithms and support vector machines. *Safety Science*, 50(4):629–644.

Zhou, Z., Bhiri, S., and Hauswirth, M. (2008). Control and data dependencies in business processes based on semantic business activities. In *Proceedings of the 10th International Conference on Information Integration and Web-Based Applications & Services*, pages 257–263. ACM.

Zhou, Z.-H. (2008). Ensemble learning. *Encyclopedia of Biometrics*.

Zicari, R. V. (2014). Big data: challenges and opportunities. *Big Data Computing*, 1:103–128.

Zikopoulos, P., Eaton, C., et al. (2011a). *Understanding Big Data: Analytics for Enterprise Class Hadoop and Streaming Data*. McGraw-Hill Osborne Media.

Zikopoulos, P., Eaton, C., et al. (2011b). *Understanding Big Data: Analytics for Enterprise Class Hadoop and Streaming Data*. McGraw-Hill Osborne Media.

Zikopoulos, P., Eaton, C., et al. (2011c). *Understanding Big Data: Analytics for Enterprise Class Hadoop and Streaming Data*. McGraw-Hill Osborne Media.

Index

For Product Safety Concerns and Information please contact our EU
representative GPSR@taylorandfrancis.com
Taylor & Francis Verlag GmbH, Kaufingerstraße 24, 80331 München, Germany